W0041979

A Short Course in Computational Science and Engineering

C++, Java and Octave Numerical Programming with Free
Software Tools

Building on his highly successful textbook on C++, David Yevick provides a
concise yet comprehensive one-stop course in three key programming
languages, C++, Java and Octave (a freeware alternative to MATLAB®).

Employing only public-domain software to ensure straightforward
implementation for all readers, this book presents a unique overview of
numerical and programming techniques relevant to scientific programming,
including object-oriented programming, elementary and advanced topics in
numerical analysis, physical system modeling, scientific graphics, software
engineering and performance issues. Relevant features of each programming
language are illustrated with short, incisive examples, and the installation and
application of the software are described in detail. Compact, transparent code in
all three programming languages is applied to the fundamental equations of
quantum mechanics, electromagnetics, mechanics and statistical mechanics.
Uncommented versions of the code that can be immediately modified and
adapted are provided online for the more involved programs.

This compact, practical text is an invaluable introduction for students in all
undergraduate- and graduate-level courses in the physical sciences or
engineering that require numerical modeling, and is also a key reference for
instructors and scientific programmers.

DAVID YEVICK is a Professor of Physics at the University of Waterloo. He has
been engaged for 30 years in scientific programming in various fields of optical
communications and solid state physics at numerous university and industrial
establishments, where he performed pioneering work on the numerical
modeling of optical communication devices and systems. Professor Yevick is
currently a Fellow of the American Physical Society, the Institute of Electrical
and Electronics Engineers and the Optical Society of America as well as a
registered Professional Engineer (Ontario). He has taught scientific and
engineering programming for over 20 years and has authored or co-authored
over 170 refereed journal articles.

A Short Course in Computational Science and Engineering

C$_{++}$, Java and Octave numerical programming with free software tools

David Yevick

CAMBRIDGE
UNIVERSITY PRESS

Shaftesbury Road, Cambridge CB2 8EA, United Kingdom

One Liberty Plaza, 20th Floor, New York, NY 10006, USA

477 Williamstown Road, Port Melbourne, VIC 3207, Australia

314–321, 3rd Floor, Plot 3, Splendor Forum, Jasola District Centre, New Delhi – 110025, India

103 Penang Road, #05–06/07, Visioncrest Commercial, Singapore 238467

Cambridge University Press is part of Cambridge University Press & Assessment, a department of the University of Cambridge.

We share the University's mission to contribute to society through the pursuit of education, learning and research at the highest international levels of excellence.

www.cambridge.org
Information on this title: www.cambridge.org/9780521116817

© David Yevick 2012

This publication is in copyright. Subject to statutory exception and to the provisions of relevant collective licensing agreements, no reproduction of any part may take place without the written permission of Cambridge University Press & Assessment.

First published 2012

A catalogue record for this publication is available from the British Library

Library of Congress Cataloging-in-Publication data
Yevick, David.
A short course in computational science and engineering : C++, Java, and Octave numerical programming with free software tools / David Yevick.
    p.   cm.
Includes index.
ISBN 978-0-521-11681-7 (hardback)
1. Computer programming – Textbooks.   2. Computer science – Textbooks.   I. Title.
QA76.6.Y48   2012
005.1 – dc23   2011044368

ISBN   978-0-521-11681-7   Hardback

Additional resources for this publication at www.cambridge.org/yevick

Cambridge University Press & Assessment has no responsibility for the persistence or accuracy of URLs for external or third-party internet websites referred to in this publication and does not guarantee that any content on such websites is, or will remain, accurate or appropriate.

אָמַר רַבִּי יוֹסֵי בֶּן קִסְמָא, פַּעַם אַחַת הָיִיתִי מְהַלֵּךְ בַּדֶּרֶךְ וּפָגַע בִּי אָדָם אֶחָד, וְנָתַן לִי שָׁלוֹם, וְהֶחֱזַרְתִּי לוֹ שָׁלוֹם, אָמַר לִי, רַבִּי מֵאֵיזֶה מָקוֹם אַתָּה, אָמַרְתִּי לוֹ, מֵעִיר גְּדוֹלָה שֶׁל חֲכָמִים וְשֶׁל סוֹפְרִים אָנִי, אָמַר לִי, רַבִּי רְצוֹנְךָ שֶׁתָּדוּר עִמָּנוּ בִּמְקוֹמֵנוּ וַאֲנִי אֶתֵּן לְךָ אֶלֶף אֲלָפִים דִּנְרֵי זָהָב וַאֲבָנִים אָמַרְתִּי לוֹ אִם אַתָּה נוֹתֵן לִי כָּל כֶּסֶף וְזָהָב וַאֲבָנִים, טוֹבוֹת וּמַרְגָּלִיּוֹת טוֹבוֹת וּמַרְגָּלִיּוֹת שֶׁבָּעוֹלָם, אֵינִי דָר אֶלָּא בִּמְקוֹם תּוֹרָה

Rabbi Yose ben Kisma said: Once I was walking on the road, when a certain man met me. He greeted me and I returned his greeting. He said to me, 'Rabbi, from what place are you?' I said to him, 'I am from a great city of scholars and sages.' He said to me, 'Rabbi, would you be willing to live with us in our place? I would give you thousands upon thousands of golden dinars, precious stones and pearls.' I replied, 'Even if you were to give me all the silver and gold, precious stones and pearls in the world, I would dwell nowhere but in a place of Torah.' (Ethics of the Fathers 6:9)

Rabbi Jose, Kismas son, berättade: En gång gick jag ut och vandrade, då mötte mig en människa, som hälsade mig, och jag besvarade hans hälsning. Han sporde mig: Rabbi, varifrån är du, och jag svarde honom: Från en stor stad, full av visa män och skriftlärda. Då sade han till mig: Rabbi, om du vill bo hos oss i vår stad, vill jag giva dig tusen gånger tusen guldmynt, ädelstenar och pärlor. Jag svarade honom: Om du så gåve mig all världens silver, guld, ädelstenar och pärlor, skulle jag aldrig vilja bo på ett annat ställe än där Torahn har sin hemvist. (Fädernas Tankespråk 6:9)

Contents

Chapter 1
Introduction

Soon after the publication of my C++ textbook, *A First Course in Computational Science and Object-Oriented Programming with C++*, in 2005, I conceived of including a yet more compact introduction to C++ in a survey of the entire field of scientific programming. Drawing on 20 years of experience of teaching programming at all levels in both physics and electrical engineering departments, I resolved to both summarize my previous treatment of C++ and incorporate a discussion of the Octave and Java programming languages, focusing on their conceptual foundations. Finally, I would insert many additional scientific programming examples, emphasizing short programs that illustrate key algorithms. By employing *only free software*, this would create a uniquely comprehensive treatment of the full set of steps from compiler installation to sophisticated scientific programming.

1.1 Objective

This textbook overviews modern scientific programming, including numerical analysis, object-oriented programming, scientific graphics, software engineering, numerical analysis and physical system modeling. Consequently, knowledge of the material will provide sufficient background to enable the reader to analyze and solve nearly all normally encountered scientific programming tasks.

1.2 Presentation

The text is concise, focusing on essential concepts. Examples are intentionally short and free of extraneous features. To promote retention, the book repeats key topics in cycles of gradually increasing difficulty. Further, since the process of learning computer language shares many similarities with that of acquiring a spoken language, *important code is highlighted in gray. Memorizing these features greatly decreases the time required to achieve proficiency in programming.*

1.3 Programming languages

Computing paradigms have evolved with time from the implementation of individual operations within a computing device to high-level structures that closely resemble interactions of physical objects with their environment. These concepts have simultaneously been realized through languages that have progressed from machine language to procedural, object-oriented and visual programming.

General-purpose procedural languages. A procedural language is composed of a structured sequence of commands, which may be further organized into modules termed functions or subroutines. A program implements a series of commands that are sequenced through logical statements. This strategy yields languages that are easily learned and applied. Especially early procedural languages such as FORTRAN, however, contain numerous unsafe constructs that invariably lead to coding errors.

Scientific procedural languages. To simplify small proof-of-principle computations, specialized scientific languages such as MATLAB® and Octave and symbolic manipulation languages such as MAPLE® or Mathematica® provide an easily learned high-level user interface to a unified built-in array of easily called and highly optimized numerical, scientific and graphical libraries. MATLAB code can be transformed into C++ through an add-on product while C++ and FORTRAN routines can be called by a MATLAB program with some effort. Additionally, MATLAB and similar programs originate from a single commercial source and therefore function nearly identically across all supported platforms (running the same version number). However, the suppression of advanced features such as classes, type-checking and user-controlled memory management can lead to structural confusion, programming errors and runtime inefficiency for larger problems. Further, the software is unavailable at many sites because of its substantial cost, although this can, however, increasingly be circumvented, through free software packages that imitate MATLAB commands. This textbook accordingly employs the most widely employed alternative, GNU Octave.

Object-oriented languages. The fundamental high-level unit in modern programming languages is an object. An object is a simplified model or *abstraction* of a particular entity. To illustrate, consider for definiteness a voltage meter. The meter has many attributes – in the extreme case the position and velocity of each atom – but only a few of these are typically of interest. These relevant attributes, which could include both user-accessible, **public**, data and behaviors, such as the voltage reading and the meter's response to depressing the power-on switch, and inaccessible, **private**, characteristics, such as the currents through individual circuit elements, compose the relevant abstraction of the object. By analogy, in a C++ program, **public** properties can be accessed throughout the code while **private** members are accessible only to functions that exist within the object itself, restricting the associated code region subject to inadvertent errors.

In an object-oriented language, objects with similar properties are described by a class. Two functions or variables with the same name that belong to different classes are considered to be unrelated, circumventing name collisions. A class can incorporate the features of a second class by *inheriting* its non-**private** properties; the new, derived, class can then employ or redefine the properties of the original, base, class without recoding. Refinements to the original class automatically propagate to the new class.

Since additional programming syntax is required in order to create and manipulate objects and classes, object-oriented languages require more time to learn than procedural languages. However, the structure of the resulting program closely represents the physical objects that are being modeled. Accordingly, object-oriented development is advisable for large programs or programs that will be frequently revised.

The C++ language. C++, which extended the preexisting C procedural programming language, constituted the first widespread object-oriented language. Numerous scientific programming packages are at present available in C++, while FORTRAN programs can be, with some effort, accessed from C++, c.f. Appendix D of my companion textbook *A First Course in Computational Science and Object-Oriented Programming with C++*. However, the additional functionality of C++ enables manipulations that can introduce unexpected dependences among variables. To ascertain these dependences, C++ typically runs more slowly than FORTRAN, although advanced C++ language features can be employed to circumvent these difficulties, as discussed in Chapter 21 of the above reference.

Java: As a more recent object-oriented language, Java provides a far broader standard feature set than C++. Classes that e.g. handle graphics and internet communications are native to the language and in principle function identically across all Java implementations (although, in reality, version and machine dependences exist). Modern programming features such as multithreading and object serialization are additionally included. However, since the language is oriented toward the corporate market, many design choices are unfavorable for sophisticated scientific programming. For example, C and C++ contain high-level commands that enable direct access to hardware resources. These include addressing and modifying the contents of individual memory locations and precisely allocating and releasing the memory available to a program during execution. Since severe errors result if such manipulations are improperly performed, Java handles such operations automatically, at the cost of longer or unpredictable execution times. Further, extensive mathematical or scientific program libraries are ported only very slowly, if at all, to new programming languages.

1.4 Language standards

As requirements evolve, programming languages undergo periodic revision by a standards organization. For relatively new languages such as Java, revisions can be significant; further, existing language elements can become deprecated (unsupported) and eventually obsolete. In contrast, revisions to mature languages such as C++ are relatively minor and do not affect core functionality. However, programs employing elements of a new standard will not necessarily function on older compilers.

1.5   Chapter summary

The organization of this book is as follows. After a short introduction to Octave in Chapter 2, the following chapters summarize the installation of a free C++ programming environment, computer hardware and software architecture and the basic structure and syntax of the C++ language. Chapter 7 introduces object-oriented programming in C++, with more advanced features of C++ following in Chapters 9–14. Chapters 15–17 discuss basic Java programming, with advanced Java features relegated to Chapter 18. Chapters 19–24 finally discuss applied numerical analysis in the context of numerous physical and engineering applications, including mechanics, electromagnetism, statistical mechanics, quantum mechanics and optics.

1.6   How to use this text

The reader is encouraged to follow the steps below.

(1) Skim through the text.
(2) Reread the chapter, programming and running as many sample programs in the text as possible. Attempt, if possible, to extend these programs.
(3) *Memorize the programs or program sections marked in gray in the text.* Success in programming is largely dependent on being able to recall instantly central language features.

1.7   Additional and alternative software packages

While comprehensive freeware and commercial C++ and Java numerical libraries exist, such as the GNU, CERN, IMSL and NAG libraries, such routines are typically designed for a restricted set of hardware and software platforms. Therefore, for smaller programs well-documented source code such as the programs in this book will often provide a more optimal trade-off between computational efficiency and development time.

Chapter 2
Octave programming

For small programs or rapid prototyping of ideas and methods, the commercial MATLAB® language, or its freeware alternatives, offers a practical alternative to C++ or FORTRAN. In this book, the free GNU Octave implementation is discussed from a scientific programming perspective. After becoming familiar with the central language constructs summarized below, the built-in Octave help facilities conveniently provide information on specialized, infrequent commands.

2.1 Obtaining octave

The Windows and Mac installation packages for GNU Octave are currently located at octave.sourceforge.net. Linux versions are available at the main Octave web site www.gnu.org/software/octave. When the program is installed, a variety of additional packages and the creation of a database of C++ components accessible by the editor can be selected. Unless space is an issue, these options should be chosen.

2.2 Command summary

(1) *Running Octave.* After clicking on the Octave icon, statements are entered interactively by typing into the resulting command window at the > prompt. An Octave session is terminated by typing **quit**.

(2) *System commands.* To change from the startup directory (folder) to the directory that either contains or will contain program files, type **cd X:\dir1\dir2\ . . . \programDirectory**, where **X** is the partition (logical drive) containing the desired directory and **\dir1\dir2 . . . \programDirectory** is an ordered sequence of the names of the directories enclosing the directory, **\programDirectory**, in which the program is located. If one or more directory names contain spaces, the entire expression containing these names must be surrounded by double apostrophes ("), e.g. **cd "X:\My Documents"**. Representative operating-system commands that can be issued from the Octave prompt include **mkdir directoryName**, which creates the directory **directoryName**, **rmdir directoryName**, which removes this directory,

dir or **ls**, which display the contents of a directory, **..**, which moves to one directory higher in the directory tree, **.**, which represents the current directory, **rename file1.1 file2.2**, which renames the file **file1.1** to the name **file2.2**, and **copy**, which similarly copies a file.

(3) *MS-DOS and Unix commands.* Standard DOS commands on Windows systems and Unix commands on Unix systems are issued in Octave by typing e.g. **dos 'copy file.1 file.2'** or, on a Unix system, **unix 'cp file.1 file.2'** (single or double apostrophe). In MATLAB such commands can also be preceded with **!**.

(4) *Command structure and continuation lines.* Octave commands end at a carriage return, comma or semicolon; however, only a semicolon suppresses the output of the statement from being written to the terminal. Two or more commands situated on the same line must be separated by commas or semicolons. A statement can span several lines but each line must normally be terminated by a three-period continuation character, **. . . .**

(5) *Creating and editing files.* The command **diary on** stores subsequent commands entered from the keyboard in a file named **diary** until **diary off** is issued. To examine this file or to create or edit an Octave program, after navigating to the directory in which the file resides, type **edit** at the command prompt, followed, where applicable, by the name of the file to be edited or created.

(6) *Comments.* Any text to the right of a comment character, **%**, constitutes a comment and is consequently ignored by the Octave interpreter. The beginning of a program should contain the date, version number, title and author. Every set of statements (a paragraph) performing a certain task should be *preceeded* by one or more blank lines followed by comment lines explaining the purpose of the program unit. When a variable is introduced, its meaning should be made clear by a comment either above the line or on the same line to the right of the statement. Such annotations insure the long-term readability of programs.

(7) *Using help.* To find and implement rarely employed commands, type first **lookfor subject** to obtain a list of all commands involving the operation 'subject'. Issuing **help commandName** (or **doc commandName**) then provides help on the command **commandName**.

(8) *Octave programs.* Octave program and function files must possess a **.m** extension; that is, to program in Octave, first type **edit** from within the Octave command window and create a file such as

```
s = 2;
s * ... % Illustrates the comment and continuation symbols
s
```

Then, when saving the file, specify **test** in the "File Name" text entry field while in the "Save File as Type" drop-down text box select MATrix LABoratory (.m). This automatically appends the correct **.m** extension to the file name. If the file is saved in a directory, e.g. **X:\testDirectory**, then, at the Octave prompt, type **cd** followed by the directory (including, if necessary, the partition name, e.g. **cd X:\testDirectory**) where **test.m** resides, press enter and then type **test**. The program **test** can also be called from within another **.m** file within the same directory.

(9) *Variable-naming conventions.* For clarity, variable names should start with a small letter, while subsequent words in the name should be capitalized, e.g. **numberOfPoints**. However, in Octave a name can represent an array of any size and number of dimensions, which can result in subtle errors. To prevent this, quantities with a row dimension can be indicated by a trailing R, those with a column dimension as C and a matrix with a trailing RC, e.g, **systemMatrixRC**. If matrices with different dimensions are present, the row and column sizes can be further specified as in **systemMatrixR4C8**. *Since Octave is not a typechecked language, the above conventions can still lead to severe and difficult-to-locate errors for some compound words such as wavefunction, which can be treated as one word in certain places and as two words (waveFunction) in others. Typing a single character incorrectly generates similar problems. These errors, however, can be immediately identified by typing* **who** *at the command line, which displays a list of all the currently defined variables. Any spelling error will then be evidenced by a variable name seemingly appearing twice in the list.*

(10) *Formatting conventions.* Every binary operator (+, − etc.) should be surrounded by spaces, but not unary operators as in $3 + -4.0$. Indentation should be employed for every set of statements that are under the logical control of a control statement such as **for, if, while**. Commas, semicolons, parentheses and braces should, where appropriate, be followed by spaces.

(11) *Program input.* To prompt the user from within a **.m** file to enter a single variable or array **x** from the keyboard, employ **x = input('user prompt ')**. A variable **y** that can later be employed in a logical control statement to branch into different program units is conveniently entered with **y = menu('Select the method', 'Method A', 'Method B', 'Method C');** which assigns the value 1, 2 or 3 to **y** according to the user selection.

(12) *Output formatting.* The **more** command pages subsequent output. To write out subsequent floating-point output with 16 digits of precision, type in **format long e**, to revert to the default 5 digits, type **format short e** or, equivalently, **format compact**.

(13) *Built-in constants and functions.* Important predefined scalar quantities are **e, pi, i** and **j**, both of which represent the unit complex number, and **eps**, the smallest number which when added to 1 gives a number different from 1 (machine epsilon). However, a major problem arises if these variables are redefined, for example, the command **i = [1, 3]** overwrites the intrinsic definition of **i**, which is not reinstated until a further command **clear i** is issued. Note that **i** and **j** are frequently employed as loop variables, *so that all loop variables should instead be labeled, for example, loop, innerLoop, outerLoop.*

(14) *Complex numbers.* A complex number is introduced as **c = 2.0 + 4.0i** and then manipulated with functions such as **real( )**, **imag( )**, **conj( )**, **norm( )**. Complex numbers are e.g. multiplied, divided and exponentiated in standard fashion either

by real or by other complex numbers. Functions such as **cos( )**, **sin( )**, **sinh( )** yield complex results when applied to complex quantities.

(15) *Loading arrays.* A variable name can represent a scalar or an array of any dimension. A row vector is introduced as

```
vR = [1 2 3 4];
```

or

```
vR = [1, 2, 3, 4];
```

while a column vector is entered, even from the keyboard, as

```
vC = [1
2
3
4];
```

or, equivalently (since a semicolon is largely equivalent to a carriage return),

```
vC = [1; 2; 3; 4];
```

With the transpose operator .' the above column vector can also be entered as

```
vC = [1 2 3 4].';
```

A matrix

$$mRC = \begin{pmatrix} 1 & 2 \\ 3 & 4 \end{pmatrix}$$

can therefore be entered in any of the following ways:

```
mRC = [1 2; 3 4];
mRC = [1 2
3 4];
vR = [1 3 2 4]; mRC = reshape( vR, 2, 2 );
vC = [1
3
2
4]; mRC = reshape( vC, 2, 2 );
```

(The order in which a vector is reshaped indicates that matrix elements with successive values of the leftmost, column, index are stored next to each other in memory.) The matrix element $(mRC)_{12}$ is subsequently accessed by **mRC(1, 2)**. Since scalars and arrays are manipulated identically, arrays with multiple dimensions are constructed from component vectors or from subarrays in the same manner as from scalar quantities, e.g. **vR4 = [[1 2] [1 2]]** yields **[1 2 1 2];**, while **mBlockRC = [mRC mRC; mRC mRC];** is a matrix of twice the dimension of **mRC**. While a vector or matrix expands dynamically as new elements are added as in **vR = [1 2]; vR(3) = 3;** this is computationally inefficient and memory should instead be preallocated through a statement such as **vR = zeros(1, n)**, which creates a row vector of **n** zero elements.

(16) *Size and length*. Octave maintains a record of the size of an array to prevent element access outside this range. Hence **mRC(1, 3)** yields an error if mRC is a 2×2 matrix. The command **size(mRC)** for an $M \times N$ matrix returns the array $[M, N]$. For a single-dimension array, **length(vR)** returns the length of the array **vR**. However, when applied to a two-dimensionsal array **length(mRC)** returns the maximum value of M and N, possibly leading to unexpected errors.

(17) *Matrix operations*. The $n \times n$ identity matrix is represented by **eye(n)** or **eye(n, n)**, while the $n \times n$ matrix with all unity elements is denoted **ones(n)** or **ones(n, n)**. If **s = 2** and **mRC = [1 2; 3 4]** as in item (15) above, then

$$s + \mathrm{mRC} = s * \mathrm{ones}\,(2,2) + \mathrm{mRC} \Rightarrow \begin{pmatrix} 3 & 4 \\ 5 & 6 \end{pmatrix}$$

while

$$s * \mathrm{eye}\,(2,\,2) + \mathrm{mRC} \Rightarrow \begin{pmatrix} 3 & 2 \\ 3 & 6 \end{pmatrix}$$

Very often, errors arise because of failure to differentiate between these. Multiplication similarly possesses different meanings depending on variable type. Multiplying or dividing a matrix **mRC** by a scalar **s** multiplies (divides) all elements of **mRC** by **s** while **mRC * mRC** symbolizes normal matrix multiplication and

$$\mathrm{mRC}. * \mathrm{mRC} \Rightarrow \begin{pmatrix} 1 & 4 \\ 9 & 16 \end{pmatrix}$$

implements component-by-component multiplication. Similarly **mRC^2** is **mRC * mRC**, while **mRC.^2** instead squares the individual elements of **mRC**. The dot operator functions analogously for other arithmetic operations such as **mRC ./ nRC**, which yields a matrix whose (i, j)th element is simply $(\mathrm{mRC})_{ij}/(\mathrm{nRC})_{ij}$. Standard functions such as **cos(mRC)** operate on the individual elements of **mRC**, here returning the matrix formed by taking the cosine of each element. *Two easily confused operations are array (matrix) transpose without complex conjugation, .', and transpose with complex conjugation, '*. Note that the simpler syntax is applied to the Hermitian conjugate operation, since this yields the standard norm of complex (as well as real) arrays. For a vector **vR = [1, 2]** with elements $(\mathrm{vR})_{ij}$, the dot or inner product without complex conjugation is given by **vR * vR.'**, while the (Kronecker) outer product **vR.' * vR** yields the 2×2 matrix with (i, j)th element $(\mathrm{vR})_i\,(\mathrm{vR})_j$, namely **[1 2; 2 4]**.

(18) *Matrix functions*. A few functions ending in the letter m such as the matrix exponential **expm** act on matrix arguments and are defined (although not implemented) through power-series expansions such as

```
expm( aRC )  =  eye( size( aRC ) )  +  aRC  +  aRC^2/2!
+ aRC^3/3! + ...
```

The determinant, trace, inverse, logarithm and square root of **aRC** are similarly given by **det(aRC)**, **trace(aRC)**, **inv(aRC)**, **logm(aRC)** and **sqrtm(aRC)**, respectively. The LU decomposition of a**RC** is expressed as **[lRC, uRC] = lu(aRC);**. The eigenvalues, arranged in ascending order, and the corresponding eigenvectors of a matrix **aRC** are placed, respectively, in the columns of the matrix **mVecRC** and the diagonal elements of **mValRC** through the call **[mVecRC, mValRC] = eig(aRC);**. Simply calling **eig(aRC)** returns a vector containing the eigenvalues in ascending order.

(19) *Solving linear equation systems.* The quotient of two matrices in Octave written as **mRC / nRC** denotes **mRC * inv(nRC)**, while **mRC \ nRC** instead represents **inv(mRC) * nRC**. Accordingly, the linear equation system **xR * mRC = yR** is solved by **xR = yR / mRC**, while **xC =mRC \ yC** if **mRC * xC = yC**.

(20) *Sparse matrices.* Operations on matrices with few non-zero elements are accelerated if the matrices are implemented as sparse matrices. The simplest procedure converts a full matrix **aRC** to a sparse matrix by **bRCsp = sparse(aRC)** (which is reversed with **aRC = full(bRCsp)**). Subsequent operations such as * and / are performed with sparse routines if all operands or arguments are sparse (except for an identity or zero matrix). **spy(mRCsp)** displays the locations of the non-zero elements of **mRCsp** while **speye(n)** is an $n \times n$ sparse identity matrix.

(21) *Random-number generation.* To compare the different versions of a program that incorporates a random-number generator, the random sequence should be the same in all versions. This is accomplished by introducing the statement **rand('state', 0)** at the beginning of each program. Uniformly distributed random numbers between 0 and 1 are generated singly with **rand** or as a multidimensional array with e.g. **rand(m, n)**. Typing **lookfor rand** displays information about functions for other random distributions.

(22) *Saving variables.* A variable **v** is stored in the text Octave file **filename** through **save filename v** and then recovered through **load filename**. This file can then be inspected or edited with any text editor. All variables present in the workspace are simultaneously saved and loaded in through the commands **save filename** and **load filename**. [In MATLAB the **save** command instead writes to a binary file **filename.mat** and will only save a variable **v** to an ASCII file **vdata** if the command **save vascii v –ascii** is instead employed. The variable is in this case recovered from the .mat file with **load vascii; v = vascii;** where only the first statement is employed if **v** is employed for the file name in place of **vascii**. However, retrieval from MATLAB binary files that store more than one variable presents difficulties since the original variable names are not stored as in **.mat** files.]

(23) *String manipulation.* A string such as **s = 'test'** is stored as an array ['t' 'e' 's' 't'] of characters so that **s(1)** returns **t** and **['a ', s]** yields the new string **'a test'**. Integers and floating-point numbers are translated into strings through the functions **int2str( )** and **num2str( )**, respectively; the reverse operation is performed by **str2int( )** and **str2num( )**. An Octave expression, expressed as a string, is sent to the command processor by writing **eval( )**; as an example, to display the files present in the directory, the commands **s = 'dir'; eval(s);** can be employed.

(24) *Cell arrays.* To form an array composed of different types of variables (including additional cell arrays) as members, write e.g. **cA** = {{**1 2; 3 4**} **'s'; 3 5**}. Then **cA** {**1,2**} is the letter s while **cA**{**1,1**}{**2,2**} is 4. In MATLAB **cellplot(cA)** additionally yields a graphical display of the contents of **cA**.

(25) *Iterators.* In Octave, **for** loops are implemented far less efficiently than vectorized expressions containing array variables. The colon operator generates a row vector, by default with unit step, that can replace iterators. For example, **1 : 3** yields the row array **[1 2 3]**, while **vR = sin(pi: -pi / 10: -1.e-4)**; produces the array **vR = [sin(pi) sin(9 pi / 10) ... sin(pi / 10) 0]**. A non-vectorized **for** loop possesses the form **for loop = 0.9 : 0.1 : 0.3** ... statements to be iterated over ... **end** or, equivalently, **for l = [0.9 : 0.1 : 0.3]** .... *A common error here is to write **for l = .9, -.1, .3** in place of one of the statements above.* This assigns the value **.9** to **l** and then executes the meaningless statements **-1** and **3** without an error message. Note that the common choice of **i** or **j** as iteration (or ordinary) variables precludes their further interpretation as complex numbers, so that **i** *and* **j** *should not be employed as variable or iterator names.* Iteration can also be implemented with a **while** loop, as in **while loop > 0.3, loop = loop - 0.1, end**, which takes the place of the **for** loop introduced at the beginning of this paragraph. A **break** statement when encountered terminates the iteration and passes control to the statement following the **end** statement of the loop in which the **break** occurs.

An isolated colon, **:**, iterates through all the rows or columns of a matrix, so that **mR2C4(:, 1) = vR2(:)** (or equivalently **= vR2.'(:)**, which is a column vector), places all the elements of the two-element row vector **vR2** into the first column of the matrix **mR2C4**. A matrix can also be mapped into a vector with the colon operator; for a matrix **mRC = [1 2; 3 4]**, writing **V = M(:)** yields the column vector **V = [1; 3; 2; 4]** (since, as noted earlier, in Octave successive *column* elements are adjacent in memory). Related constructs are **linspace(S1, S2, N)** (and **logspace**), which generate **N** equally (logarithmically) spaced points between **S1** and **S2** that include the endpoints.

(26) *Control logic.* The logical operators in Octave are given by ==, <, >, >=, <=, ~= (not equal) and the and, or and not operators **&**, | and ~, respectively. These are typically employed in **while ... end, if ... elseif ... end** (note that **elseif** is a single word) and **switch** statements. The last of these is rather complex, so the help subsystem should be consulted when coding.

(27) *Functions.* A function, with a name **functionName**, must be placed in a separate (and similarly capitalized) file **functionName.m** (this requirement can be circumvented somewhat in Octave). The first line of this file must further be of the form **[mOutRC, vOutR, ...] = functionName(mInRC, vInR, ...)**. To share internal variables with other program units a **global** statement that includes the names of the shared variables separated by spaces (not commas) is introduced as follows (the **endfunction** statement is generally not required; and a helpful convention is to capitalize global variables):

```
%file functionName.m:

function [vOutR, xOut] = functionName( vInR, xIn )
global G ICONV
G = length( vInR );
vOutR = [vInR(1), vInR(2)^2];
xOut = 0;
ICONV = vInR(1) * 3;
endfunction

%program file test.m.
%Only G is visible and shared from within this program unit:
global G
yInR = [1 1];
[yOutR, yOut] = functionName( yInR, 2 );
G
```

(28) *Important functions.* Some often-employed functions include **rem(n, m)**, the remainder of n/m, which is positive or zero for $n > 0$ and negative or zero for $n < 0$, **ceil( )**, **floor( )**, **fix( )** and **round( )**, which round a real number up and down, toward zero and toward the nearest integer, respectively. Common functions with vector arguments encompass the discrete forward and inverse Fourier transforms, **fft( )** and **ifft( )**, and **mean( )**, **sum( )**, **min( )**, **max( )**, and **sort( )**. Interpolation of data is performed by **y1 = interp1(x, y, x1, 'method')**, where 'method' is **'linear'** (default), **'spline'** or **'cubic'**, **x** and **y** are the input x- and y-coordinate vectors and the scalar or vector **x1** contains the x-coordinate(s) of the point(s) at which the interpolated values are desired. The function **roots([1 3 5])** returns the roots of the polynomial $x^2 + 3x + 5$.

(29) *Function arguments.* Numerous functions accept other functions as arguments. These can be entered in different ways, e.g. **fzero('cos', 2)**, which returns the zero of the function **cos** closest to 2. The use of apostrophes, however, is here restricted to Octave functions; a more general syntax is **fzero(@cos, 2)** or, equivalently, **aF = @cos; fzero(aF, 2)**. In these statements @ denotes a *function handle* with the property that **aF(0.1)** evaluates to **cos(0.1)**. Some important Octave functions of this type are **min('functionname', a, b)**, which searches for a minimum between the lower and upper endpoints **a** and **b**, and **fmins('functionname', x)**, which searches for a root in two-dimensions closest to the vector **x**. As well, **fzero('functionname', a)** finds the zero closest to **a**, **quad('functionname', a, b, tol)** integrates a function from **a** to **b** to within an error **tol**, and **lsode( )** (in Octave) and **ode23( )** or **ode45( )** (in MATLAB) solve systems of coupled first-order ordinary differential equations (ODEs).

(30) *Graphic operations.* Calling **plot(x1R, y1R, x2C, y2C, 'r', ...)** with column or row vectors as arguments plots multiple sets of (x, y) curves, the first of which is blue by default and the second of which is here specified as **'r'**, red. Executing **plot(cxR, 'g∗')**, where the elements of **cxR** are complex, instead graphs the imaginary part against the real part of these elements with green stars. Logarithmic graphs

are similarly plotted with **semilogy( )**, **semilogx( )** or **loglog( )**. Three-dimensional grid and contour plots are created with **mesh(mRC)** or **mesh(xR, yR, mRC)** and **contour(mRC)** or **contour(xR, yR, mRC)**, where **xR** and **yR** are vectors (either row or column) that contain the x and y positions of the grid points. A plot generated in a program (**.m**) file is not necessarily displayed until a **drawnow**, **pause(seconds)** or (normally) a subsequent **plot** statement is reached. The command **hold on** (cancelled by **hold off**) prevents the erasure of the graph window so that additional curves can be overlaid.

The figure window is divided into an $m \times n$ two- dimensional array of subplots of which the pth is selected for the next plot by the command **subplot(m, n, p)**. Here $p = 1$ corresponds to the row 1, column 1 plot position, while $p = 2$ refers to the row 1, column 2 plot position and so on. The command **clf** clears the figure window, while **figure(n)** for integer n opens a new graphics window called Figure n. Functions are conveniently plotted with the command **fplot('functionname', [a b])**, where **a** and **b** are the initial and final x-values. Axis defaults are overridden with **axis([xMinR xMaxR yMinR yMaxR])**. By placing additional optional arguments into the **axis** command the most common graphical functions can be manipulated. Axes are also labeled with **xlabel('xtext')** and **ylabel('ytext')**. Commands such as **print –deps** and **print –dtiff** yield encapsulated postscript and TIFF graphics files of the current plot window, respectively (**help print** displays all options), while **orient landscape** changes the printer or graphics output to landscape format. To access and change all the properties of the graph, first enter **get(gca)** to see a list of the graphics properties and their settings. Then employ e.g. **set(gca, 'linewidth', 10)** to change a selected property.

(31) *Memory management.* Since user-defined variable names hide built-in variable and function names, if a program defines a variable such as **length**, the built-in function **length(x)** ceases to function. The user-defined variable can, however, be destroyed through the command **clear length**. Each program should therefore begin with **clear all** to eliminate all preexisting variables. *This especially prevents subtle errors generated by resizing variables defined during a previous run*; otherwise, if for example, an array has a size of 100 the first time the program is run, then, if 50 elements are written into it the next time the program is edited and reexecuted, the remaining 50 elements are still allocated and contain the previous data. The inner product of this vector with itself will therefore erroneously include contributions from these values. After memory has been deallocated with the **clear** command, the memory space assigned to the program is not decreased until garbage collection is activated through the **pack** command.

(32) *Structures.* A structure in Octave functions as a generalized array that segregates variables of different types under a single heading. This prevents collisions between variables that have the same name but refer to different types of objects. In Octave a structure can be employed simply by appending the member-of operator (.) to the structure name, as in

```
Spring1.position = 0;
Spring1.velocity = 1;
Spring1.material = 'Steel';
Spring1.position = Spring1.position + deltaTime ...
        k / m * Spring1.velocity
```

2.3   Logistic map

As a simple example of an Octave program, we implement the *logistic map* defined by the recursion $x_{n+1} = \mu x_n(1 - x_n)$ for $0 < x_0 < 1$. The x_n tend to a single value as n is increased for $\mu < 3$ but fluctuate among multiple values and subsequently exhibit period doubling into chaos as μ is increased from 3 to 4. That is, in the $\mu < 3$ regime, any initial value reaches a fixed point $x_* = (\mu - 1)/\mu$ as the number of iterations becomes large, while for μ slightly above 3 the system instead oscillates between two fixed values. The number of these values subsequently increases with μ and eventually becomes infinite. In this chaotic regime, infinitesimal changes in the starting value generate unpredictable variations in the output after a large number of iterations, as can be verified by running the program below for μ equal to e.g. 2.6, 3.3, 3.52 and 3.9:

```
numberOfBins = 100;
numberOfSteps = 100;
growthConstant = 3.6;
startValue = 0.5;
histogramR = zeros( 1, numberOfBins );
mapValuesR = zeros( 1, numberOfSteps );
mapValuesR( 1 ) = startValue;
histogramR( mapValuesR( 1 ) * numberOfBins + 1 ) = ...
    histogramR( mapValuesR( 1 ) * numberOfBins + 1 ) + 1;
for loop = 2 : numberOfSteps
    mapValuesR( loop ) = growthConstant * mapValuesR ...
        ( loop - 1 ) * ( 1 - mapValuesR( loop - 1 ) );
    histogramR( ceil( mapValuesR( loop ) * numberOfBins ) ) = ...
    histogramR( ceil ( mapValuesR( loop ) * numberOfBins ) ) + 1;
end
figure( 1 )
plot( mapValuesR );
figure( 2 )
plot( histogramR );
```

Chapter 3
Installing and running the Dev-C++ programming environment

We commence our discussion of C++ by describing the installation process for the free Dev-C++ development environment and DISLIN graphics software. Dev-C++, which can be downloaded from http://www.bloodshed.net/dev/devcpp. html, offers a simply managed, integrated set of development tools based on the Linux g++ compiler both on the native Linux platform and as a Windows program. The platform contains, among other programs, a FORTRAN 77 compiler (g77), a debugger (gdb) and a profiler (gprof). The DEV-C++ version described here is the 4.9.9.2 version 5 beta. For instructions relevant to previous versions, consult *A First Course in Computational Science and Object-Oriented Programming with C++*. When DEV-C++ is installed and executed for the first time, a window appears with language and directory preferences, followed by a second window asking whether a class list should be built. Answer "Yes" in the second window to increase the versatility of the editor.

3.1 Compiling and running a first program

(a) *Entering a program*. Two simple methods exist for creating and executing a new program. The first of these generates a *project*, which implies that files that are later placed in the project will be subsequently processed together by the environment. To create a project, double click on the DEV-C++ icon and in the program window depress the button marked "New" on the button bar. From the drop-down menu that appears select Project and then in the pop-up window select Console application, enter an appropriate name for the project, insure that the radio button labeled C++ project is selected and press OK. In the subsequent secondary window entitled Create new project, specify the directory where the files should be placed and name the project. An editor window appears containing the skeleton code

```
#include <cstdlib>
#include <iostream>
using namespace std;

int main(int argc, char *argv[])
{
  system("PAUSE");
```

```
    return EXIT_SUCCESS
}
```

and places the cursor at the beginning position of the window. Using the tab key for indentation, add the additional line

```
cout << "Hello World" << endl;
```

immediately after the first opening brace, {. The effect of **cout** << is to display the quantity to the right of the *insertion operator* << on the terminal, while << **endl** terminates the output line. Also add the following comment lines at the beginning of the program:

```
// Hello world v. 1.0
// Aug. 11 2000
// (your name)
// This program tests the C++ environment
```

Any text on a line to the right of two forward slashes is treated as a comment and is not processed as part of the C++ program. *The lines before* **int main(...)** *together with the first two of the three last lines of the program will not be repeated in most of the remaining programs in this textbook, although they will be present in every console program that is created by Dev-C++. The initial statements can normally be replaced by a single statement* **#include <iostream.h>**, *which is, however, an antiquated programming strategy.*

To compile and run a single file without first creating a project, after opening Dev-C++ depress the third button marked "New" on the button bar in the program window. The program above can now be entered or, alternatively,

```
// Hello world v. 1.0
// Apr. 18 2011
// (your name)
// This program tests the C++ environment
#include <iostream>
#include <windows.h>
using namespace std;

main ( ) {
  cout << "Hello World" << endl;
  Sleep(4000);
}
```

Another variant, which as in the first program of this section pauses the program indefinitely until any key is depressed:

```
#include <iostream >
#include <conio.h>
using namespace std;

main ( ) {
  cout << "Hello World" << endl;
  cout << "Press any key to continue" << endl;
  getch( );
}
```

Finally, the simplest procedure, and the easiest to remember:

```
#include <iostream >
using namespace std;

main ( ) {
    cout << "Hello World" << endl;
    cout << "Press any key to continue " << endl;
    int endProgram;
    cin >> endProgram;
}
```

In entering programs *names are frequently misspelled or improperly capitalized*, leading to unexpected consequences, especially when similar characters are substituted for one another. In particular 0 and O and 1 and l are difficult to distinguish at low screen or printer resolution and can lead to errors such as when 1 < 10 (1.0 < 10) is employed in place of l < 10 (the letter l < 10). *As well, the semicolon at the end of a statement is often omitted* and two successive lines are read by the compiler as a single statement. This often produces cryptic error messages with incorrect line numbering. In general, *fine details of the program syntax*, such as double quotation marks rather than single quotation marks and braces instead of parentheses should be carefully noted.

(b) *Saving the file and running the program.* Now click on the third icon from the left on the lower button bar with float-over text "Compile and Run" and type in a name for the file (do *not* add a .cpp extension to your file name here – this is performed automatically). A compilation progress window should appear, after which the program executes.

(c) *Opening a preexisting file.* To open a .cpp file in Dev-C++ click on the second icon from the left on the upper button bar with the associated float-over text "Open Project or File" and select the desired icon from the secondary File window. After editing the file, selecting the compile and run icon will automatically save, process and execute the new code.

3.2 The Dev-C++ debugger

Often the most efficient procedure for locating errors that arise when a program is executed is to inspect the values of the variables in the program during execution. Often this is best accomplished simply by adding additional code lines that write these values to the terminal or to a file, but variables can also be examined during program execution through a debugger. In this section, an error is introduced into our Hello World program in order to demonstrate both procedures.

Return to one of the programs you created in the section entitled "Creating and Running a First Program" and incorporate both a programming error and debugging lines into the program as follows:

```
int main(int argc, char *argv[])                     // or simply main( )
```

```
{
    int i = 0;
    cout << "The value of i is " << i << endl;
    i = 5;
    cout << "The value of i is " << i << endl;
    int j = 0;
    cout << "The value of j is " << j << endl;
    int k = i / j;
    cout << k << endl;
}
```

Save this file by clicking on the fourth icon on the top button bar that resembles a floppy disk with the float-over text "Save".

To activate or deactivate debugging, select "Tools → Compiler Options" from the menu bar, followed by the Settings tab (the arrow indicates that the "Tools" menu option should be selected, followed by the "Compiler Options" submenu item). Click on "linker" in the left windowpane; "Generate Debugging Information" appears as a text label on the right-hand pane. Clicking on "Yes" or "No" to the right of this text generates a drop-down text box from which the desired behavior can be selected. Profiling, which details the amount of time spent in different code regions, can similarly be activated by selecting "Code Profiling" in the left-hand windowpane. Increasing the compiler optimization level through the menu option "Optimization" often reduces the execution time of the program, but should generally be attempted only at the end of the development process.

With debugging activated, recompile the program by selecting the first icon in the lower toolbar. Inside the editor window, click in the gray area just to the left of the line **int i = 0;**. A check symbol should appear and the line should turn red; this is termed a *breakpoint*. Click again on the Debug icon. A set of debugging menu items will appear at the bottom of the editor window. Locate the windowpane to the left of the main editor window and right click inside this area. A pop-up menu with the selection "Add Watch" will appear (if this fails simply click on the "Add Watch" icon in the debugging toolbar). Click on this icon. A secondary window requesting a variable name appears. Type in **i** and depress the OK pushbutton. An icon labeled **i** becomes visible in the left-hand windowpane. Depress the "Run to Cursor" icon in the debug toolbar. The active line advances to the first breakpoint. Now identify the upper left-hand icon in the debug toolbar labeled "Next Step" and select this icon repeatedly. The active line, marked in blue, will move through the program synchronously as the value of the variable **i** is seen to change. When the position of the error due to the invalid division by zero is reached, the active line cannot be further updated. Of course, in this example the error can also be identified from the data displayed on the terminal through the **cout** lines. For future reference, the Step Into icon is employed to enter a function. That is, to step through lines within a function (change the scope to the function), the Step Into icon should be selected once the active line has been located at the position of the function call.

3.3 Installing DISLIN

DISLIN provides high-performance professional scientific graphics routines that can be employed within numerous C, C++ and FORTRAN programming environments. The DISLIN package is available free of charge for Dev-C++ and other non-commercial compilers at http://www.linmpi.mpg.de/dislin/. The installation steps for DISLIN are as follows.

(a) Download from the DISLIN website http://www.mps.mpg.de/dislin/windows.html the DISLIN distribution dl_**_mg.zip for GCC, where ** represents the current version number, and then unzip and install the program. Be sure to set when prompted for a directory name the drive letter to the same letter X: as was employed in the Dev-C++ installation, while retaining the remainder of the default directory name. The software will then be installed in the directory X:\dislin.

(b) From the start menu, select All Programs → Bloodshed Dev-C++ → DevC++ to start the program. Select Tools → Compiler Options from the menu bar and select the second check box entitled "Add these commands to the linker command line" and then click on the checkbox so that a check appears. In the text-entry field below this line enter

```
X:\dislin\dismgc.a -luser32 -lgdi32 -lopengl32
```

Now select the tab entitled "Directories" and then the subtab "C++ Includes". At the bottom of the list of directories enter

```
X:\dislin
```

Finally, depress the OK pushbutton.

3.4 A first graphics program

A sample graphics program illustrates the code lines required by all DISLIN programs. Type the following into the editor window after depressing the "New" pushbutton on the right-hand side of the lower toolbar:

```
#include <iostream>        // Required by DISLIN!
#include <dislin.h>        // Required by DISLIN! Includes the
                           // plotting package
using namespace std;

int main(int argc, char *argv[])
{
        int numberOfPoints = 2;
        float x[2] = {0, 1};
        float y[2] = {0, 2};
        qplot(x, y, numberOfPoints);
}
```

It is extremely important to observe that the command **#include <iostream>** *and* **using namespace std;** *(or, alternatively,* **#include <iostream.h>***) must be* *present for DISLIN to function properly on a C++ file*. Now select the third "Compile and Run" icon from the left on the bottom button bar. Type in a suitable name for the file (again do not enter the .cpp extension). A graph of the two points should appear.

You can generate, among many other options, a TIFF, Adobe PDF or postscript file in place of the screen plot by placing one of the lines

```
metafl("TIFF");
metafl("PDF");
metafl("POST");
```

respectively, into the **main( )** program before the line containing **qplot**. A file named **dislin.xxx**, where xxx is respectively tif, pdf or eps, is then placed in your directory when the program is executed. If e.g. an .eps file **dislin.eps** already exists, the new .eps file will instead be **dislin_1.eps**, etc.

The contents of any window can also be printed by clicking the left mouse button anywhere inside the window to make it active and subsequently depressing the Print Screen key while holding down the ALT key. (Using the CTRL key instead of the ALT key instead captures the contents of the entire screen.) You can then open an application program that accepts graphics such as Paint (Start → Programs → Accessories → Paint) or an appropriate word processor and select Edit->Paste from its menu bar to insert a bitmap of the captured window that can subsequently be printed through the application program's print function.

3.5 The help system

Dev-C++ includes an abridged help system that is accessed from the menu bar. The first section of the help menu describes the operation of Dev-C++, including debugging and compiling, while the second section overviews C++.

3.6 Example

The following code calculates the magnitude of the gravitational field of a point particle both inside and outside the Earth and displays the result as a contour plot, a color graph or a three-dimensional line plot:

```
#include <iostream.h>
#include "dislin.h"
const double KM = 1000;
const double GRAVITATIONALCONSTANT = 6.67e-11;
const double EARTHMASS = 5.97e24;
const double EARTHRADIUS = 6380 * KM;
const int MATSIZE = 20;                 // Must be a const int
double gravitationalField( double aX, double aY ) {
        double radius = sqrt( aX * aX + aY * aY )
        if ( radius <= EARTHRADIUS )
                return GRAVITATIONALCONSTANT *
```

```
                  EARTHMASS * radius /
                  ( EARTHRADIUS * EARTHRADIUS * EARTHRADIUS );
   else
      return GRAVITATIONALCONSTANT * EARTHMASS /
                  ( aX * aX + aY * aY );
}

main ( ) {
   double position[matSize];         // x and y coordinate positions
   float field[matSize][matSize];    // gravitational field
   float offset = matSize / 2 - 0.5; // starting grid point
   for ( int loop = 0; loop < matSize; loop++ ) {
      position[loop] = 0.1 * earthRadius * (loop - offset);
   }
   float x, y;
   for (int outerLoop = 0; outerLoop < matSize; outerLoop++) {
      x = position[outerLoop];
      for ( int innerLoop = 0; innerLoop < matSize; innerLoop++ ) {
         y = position[innerLoop];
         field[outerLoop][innerLoop] = gravitationalField( x, y );
      }
   }
   metafl( "XWIN" );
   disini( );         // required for 3-dimensional plots
   int iPlot = 2;     // set to 1 for surface plot, 2 for color plot.
   if ( iPlot == 1 ) // surface plot
      qplsur( (float*) field, matSize, matSize );
   else if (iPlot == 2)  // color plot
      qplclr( (float*) field, matSize, matSize );
   else {              // contour plot
      int numberOfContours = 30;
      qplcon( (float*) field, matSize, matSize, numberOfContours );
   }
}
```

Note that DISLIN routines require **float** arrays as data arguments. The syntax
(**float** *) casts (converts) the subsequent variable to a float array (more precisely,
pointer) type.

Chapter 4
Introduction to computer and software architecture

Scientific programming comprises four basic elements: analyzing a physical problem, developing a numerical algorithm for solving the problem, designing a program that implements this solution within a clear and understandable framework and, finally, determining the accuracy and the limits of validity of the numerical solution. Before discussing these, however, we survey computational methods and software and hardware architecture.

4.1 Computational methods

While most textbook problems possess a high degree of symmetry and/or a limited number of variables and can therefore be solved analytically, real-world applications typically require numerical analysis. Further, even analytic solutions can be unstable, as in water flowing through a cylindrical tube at high velocity, for which the motion depends critically on the initial conditions. Numerical calculations are then performed for numerous initial conditions and statistical properties derived from the results.

Often a numerical model of a continuous system recasts the solution to the full, global problem as a set of simplified, coupled local problems, each of which describes the system over a restricted spatial or temporal domain. Continuous operators such as derivatives and integrals are then approximated by employing their definition as the limit of discrete differences and sums but without passing to the infinitesimal limit. As an example, the response of a building to an applied force from the ground can be obtained by noting that the forces on each brick vary only slightly over the surface of the brick. The response of each discrete, rigid, brick to these local applied forces can therefore easily be evaluated. Coupling the forces and displacements on each brick to those on neighboring bricks and to those applied at the boundary between the house and the ground generates a large system of linear equations that is solved numerically.

4.2 Hardware architecture

The basic principles of computer architecture are evidenced by considering a pocket calculator. In such devices, data such as numbers entered from the keypad or the results of intermediate calculations are stored either in a series of capacitors on the CPU chip or in an external dedicated memory chip. Subsequently, when other calculator buttons are pressed, an integrated circuit termed the central processing unit (CPU) retrieves the data, applies a sequence of basic instructions and sends the output to a LCD display.

A computer differs from a calculator principally in the extent of the CPU instruction set and the number of attached components that perform specialized functions. For example, arithmetic functions may be processed either in a dedicated area of the CPU circuitry called the arithmetic logic unit (ALU) or by a separate chip. Input or output is directed to devices such as a graphics card, printer or hard disk, each of which may have additional processing circuitry. However, in all cases each target memory location or device must possess a unique memory address. The CPU transfers data to or from a specific location by preceding the data by the corresponding address. This digital signal is placed as a packet onto a series of wires called the system bus. Each external device is attached to the bus wires. Dedicated circuits intercept only the packets that contain the device address.

Computational time can be decreased by rewriting mathematical or data operations so that they directly map to the CPU instruction set or by ensuring that the program primarily accesses rapid memory components. The fastest of these devices include *memory registers* within the CPU with very small latency time (additional clock cycles) for data storage. *Cache memory* inside the CPU or next to the CPU is accessed through a dedicated memory bus (the L1, or level 1, cache is physically closest to the memory, while other levels, labeled L2, L3, . . . , are larger and slower). *Random-access memory* (RAM) denotes rows of chips, located in sockets on the motherboard, that access the CPU through the memory bus. *Read-only memory* (ROM) refers to a specialized, flash, memory chip that retains its information when the computer is shut down. This chip contains the initial program required to activate the basic functions of the computer upon startup (boot) and is reprogrammed by applying voltages to the memory elements for a certain time to introduce new functionality to the hardware. Finally, *distributed memory* is located on remote networked machines and typically accessed through internet addresses.

A modern "virtual" operating system allocates memory dynamically as a program executes such that, if all fast cache and RAM memory is allocated to the programs running on the computer, blocks of data in memory that are not immediately required by the operating system are reallocated to free space on the hard disk or other physical storage devices. This operation, termed a page fault, severely retards program execution.

4.3   Software architecture

While the hardware of a computer defines its ultimate performance limits, software determines the degree to which these limits are reached. Different programming languages effect different trade-offs between performance and user accessibility as they have evolved from direct manipulation of the CPU and memory to high-level statements that closely resemble the physical objects to be modeled. Modern compilers or interpreters map complex language statements onto the underlying CPU instruction set, while routines that are frequently used or that interact directly with external devices such as sensors or measurement instruments are often still coded through low-level non-compiled instructions. The most primitive of these are termed *machine language*, which describes system-specific instructions that map directly to CPU instructions. Here a set of bits will be, for example, divided into an op code that directs the processor to, for example, copy the contents of one memory register to a second register, followed by additional sets of bits that specify the source and target register for this operation.

To avoid the obvious complexity of machine-language programming, *assembly language* instead represents a CPU instruction as a three-letter mnemonic. These mnemonics are translated into an *object file* containing a binary representation of the CPU commands by the *assembler*. The *linker* combines separate object files that may originate from different sources, including compiled high-level languages, into a single *executable program*. An object file normally possesses an .o or .obj extension, while an executable file typically is associated with an .exe extension in Windows and an .out extension or no extension in UNIX and Linux.

Finally, the syntax of a *high-level language* approximates natural language. *Procedural languages* initially formed a program from a logically sequenced set of statements and procedures such as functions and subroutines that transform input into output data. *Object-oriented* programming languages instead model the interactions between the underlying physical or abstract objects so that a program describes the object properties and their behavior as time evolves. For example, consider entering a value such as 0.5 into a calculator, depressing the button that calculates the sine of this value and, finally, selecting a button that displays the new value in memory register on the calculator screen. This sequence of events can be represented in C++ as

```
Calculator MyCalculator;
MyCalculator.inputValue( 0.5 );
MyCalculator.depressSineButton( );
MyCalculator.displayValue( );
```

While the corresponding procedural code,

```
double calculatorValue = 0.5;
double outputValue = sin( calculatorValue );
cout << outputValue << endl;
```

is far shorter, the object construct organizes the properties and functions of a physical object into a single self-contained unit that is easily modified and transferred between programs. Object-oriented languages additionally provide a foundation for still higher-level programming idioms such as graphical programming, for which right-clicking on a calculator icon reveals a list of its attributes, which in this case could be the value, **calculatorValue**, held in the calculator's internal memory register and the **inputValue( )**, **depressSineButton( )** and **displayValue( )** functions. Other icons represent objects or graphical user-interface parts that obtain data from or display data to the user. Drawing a line between two objects enables one object to call a function in the second object.

4.4 The operating system and application software

The *operating system*, such as Windows®, Linux or UNIX, interfaces the application programs to the computer hardware. An application therefore generally obtains hardware resources by calling operating-system functions. A modern multitasking operating system services these requests in a manner consistent with the relative priorities of other currently running programs. This prevents, for example, lockups resulting from contention between two programs for the same resource.

Chapter 5
Fundamental concepts

Before discussing the C++ language in detail, we summarize its conceptual foundations. Since C++ consistently adheres to a small number of basic principles, acquiring a working understanding of these greatly hastens learning of C++ syntax.

5.1 Overview of program structure

The smallest unit of a C++ program is a *token*, which is a letter or symbol that the compiler can process. Appropriate groups of tokens yield words. Of these, *identifiers* (variable names) form *atomic* (single-element) expressions. *Operators* combine expressions to form new expressions. Terminating a valid expression with a semicolon yields a *statement*, which is equivalent to a sentence. A *block*, which is analogous to a paragraph, unifies and isolates one or more statements from the remainder of the program. *Control structures* determine the program flow according to the outcome of logical operations, while *functions* and possibly *subroutines* modularize compound statements by associating a label with a frequently occurring sequence of statements. Finally, *classes* and *objects* structure related variables and functions into generalized arrays. These, like a chapter, describe a single topic, namely the properties of a related group of objects.

5.2 Tokens, names and keywords

The C++ compiler processes source-code lines in order of appearance and text within a line from left to right. A program is read as a sequence of tokens separated by non-printing *whitespace* characters, which include tabs, carriage returns and spaces. Valid tokens are a–z, A–Z, 0–9 and certain punctuation characters. Upper- and lower-case letters represent different tokens such that myVariable and myvariable are unrelated names. A word is a sequence of tokens terminated by whitespace. A *reserved keyword* is a word such as **int** or **if** for which the complier has a special interpretation.

5.3 Expressions and statements

A valid C++ expression terminated by a semicolon yields a *statement* (in contrast, a *preprocessor directive* that instead begins with the symbol #, is processed before any statements are read by the compiler, while text to the right of a comment delimiter // or between /* and */ is ignored by the compiler). A statement is composed of one or more valid *expressions* separated by *operators* such as +, −, * or /. As an example, **13;** and **aI;** represent valid statements. While these statements would normally be optimized away by the compiler, when they are joined by operators meaningful compound statements such as **aI = aI + 13;** can be generated. Since multiple whitespace characters appearing in an expression are compiled as a single space, a line can contain several statements, while statements may span any number of lines or tab characters,

```
int j = 10; int J     = 20;   int
k = 30;
```

The exception is a *string*, which consists of a quantity of text that can be entered in more than one line only if the *continuation character*, \, is present, as in

```
char s[100] = "This is \
a string"
```

A statement is a control point in that the compiler analyzes a statement only after all previous statements have been processed.

5.4 Constants, variables and identifiers

A C++ constant, such as 2.0, is a fixed *value* that can be placed in one or more memory locations only when a program is executed, since it does not have a dedicated storage location. *A variable, **int j**, on the other hand, corresponds to a physical storage location, i.e. memory space*, that is reserved by the operating system at runtime. The value at this location can in general change during program execution. The name, or *identifier*, of a variable is any continuous stream of valid tokens that does not begin with a number and does not contain punctuation marks except for the underscore character (_). However, since the names of many variables employed by the operating system start with an underscore, many programmers avoid this practice.

5.5 Constant and variable types

To utilize a variable the compiler must be informed through a *type* specification of the amount of required memory space and the interpretation of the value stored in this memory. To illustrate a sequence of bits in memory is interpreted as the letter A if typed as a **char** and as the integer 64 if typed as an **int**. Thus,

```
int m = 1;
m = m + 1;
```

requires the keyword **int** to indicate that the storage space corresponding to **m** stores integer values so that the compiler can establish that 1 can be meaningfully inserted into this memory space. Removing the **int** keyword yields a compile-time error.

The most frequently occurring variable types can be distinguished first by the number of bytes each employs for storage. Here a byte refers to a unit consisting of eight binary memory values or bits that are either zero or one. Therefore, a byte can possess $2^8 = 256$ different values. The number of bytes reserved for variable types such as **int** depends on the computer and compiler. In earlier versions of C++ created for 16-bit machines, an **int** occupies two bytes, limiting the number of possible **int** values to 65,536. (The actual storage size of a variable **m** in bytes can be determined at runtime by calling **sizeof(m)**.) In a 32-bit machine, memory is accessed through a bus of 32 wires that simultaneously send or retrieve four bytes and therefore can (optimally) address $2^{32} = 4,294,967,296$ different memory locations. An **int** variable is conveniently stored in four bytes on such machines and can then acquire $2^{32} = 4,294,967,296$ possible values that are mapped to integers from $-2,147,483,647$ to $2,147,483,648$. Although an integer between these limits is represented exactly, incrementing the largest allowable **int** by one generates the smallest allowable **int**, while decrementing the smallest allowable **int** by one similarly yields the largest **int**.

Floating-point numbers are represented in scientific notation with an accuracy of approximately 7 digits for a **float** and 14 digits for a **double**. The first bits in the variable's memory store the mantissa (the 7 or 14 significant digits) and the last few bits store the exponent. Such a representation, unlike that of an **int**, is inexact but spans a large range of magnitudes, up to $\approx 10^{\pm 38}$ for a **float** and $\approx 10^{\pm 308}$ for a **double** (the exact values can be found by including the header file **<float.h>** (or **<limits>**) and introducing the statement **cout << FLT_MAX << " " << DBL_MAX << endl;**). In most C++ compilers both **float** and **double** reserve eight bytes of storage space. Consequently the **double** keyword is generally preferred because **float** variables are often in any case inefficiently stored as **double** values after setting the least-significant bits to zero. A **double** constant is distinguished through use of the decimal point and can also be written in exponential notation by appending the suffix **e** or **E** followed by the mantissa. That is, **3e-1** represents the same **double** constant as **3.0e-1** or **0.3**. Twice the normal amount of memory is allocated to an **int** or **double** variable if its definition statement is prefixed with the **long** keyword.

Table 5.1

Integer value	Character constant	Description
0	'\0'	null character
010	'\n'	line feed
032	' '	space
048	'0'	0
049	'1'	1
. . .		
057	'9'	9
065	'A'	A
066	'B'	B
. . .		
090	'Z'	Z
097	'a'	a
. . .		
122	'z'	z

A **char** stores a single byte of data that is interpreted at runtime as the code for a single character (letter). That is,

```
char aC = 'a';
```

yields the output **a** when the following line is executed:

```
cout << aC << endl;
```

Standard single-byte **char** variables can store $2^8 = 256$ possible values that correspond to the elements of the ASCII character set. The first 32 ASCII characters are exclusively non-printing control characters such as backspace, bell, tab, etc. The most important ASCII values are shown in Table 5.1. Thus

```
char c;
cout << (c = 65) << (c = 10) << (c = 97) << endl ;
```

yields the output

```
A
a
```

A character such as **'a'** should not be confused with the character *string* **"a"**, which is a two-element array consisting of the single character 'a' followed by a byte with all of its component bits equal to zero termed the *null character*. The presence of a null character enables functions of string arguments to determine when a string terminates and thus to stop processing its bit pattern. As noted earlier, if a string is broken across several lines, a continuation character given by a backslash (\ ) must be placed at the end of each line.

5.6 Block structure

In C++, related statements are generally structured and thus afforded a degree of isolation through insertion into separate program units termed *blocks*. A block is normally signified by a set of braces, { }, and can be viewed as a compound statement or a region from which previously defined objects outside the region can be viewed and accessed but which prohibits access, except in specialized cases, to its contents from this outer region. Consequently, if a variable **i** is defined outside the block and a second **i** is defined inside, the two variables are independent and do not collide. The identifier **i** inside the block instead refers to the memory space of the second **i**, *hiding* the first variable, until the block terminates or the new variable is otherwise destroyed, as is apparent from

```
int M = 3;

int main( int argc, char *argv[ ] )
{                               // first block
    int n = 4;          '       // m is 3 and n is 4
    {                           // second block
        M = 10;
        int k = 5;              // M is 10, n is 4 and k is 5
        double n = 6.0; // M is 10, n is 6.0 and k is 5
    }
    int m = 3;                  // M is 10, n is 4, m is 3 and k no
                                // longer exists
}
```

Since **M = 10;** is not a definition, unlike **double n = 6.0;**, a new memory location is *not* allocated for **M** within the innermost block.

A variable such as **M** in the program above that is defined outside all blocks (and is conventionally capitalized) is termed a *global variable*. Global variables, even if hidden by another variable of the same name, can be accessed throughout the program through the syntax **::M**, where **::** is termed the *scope-resolution operator*.

To summarize, suppose an outer region R contains an inner block B. Then the following conditions apply.

- Variables previously defined in R can be accessed in B.
- Variables defined in B are destroyed (vanish) when B is exited.
- If a variable is defined in B with the same name as a (non-global) variable in R, the variable in R subsequently becomes inaccessible in B. The type of the new variable can also differ from that of the hidden variable because the two storage locations are independent.

5.7 Declarations, definitions and scope

Statements that exclusively indicate to the compiler the type and extent in memory of a variable (*declarations*) must be distinguished from those that lead to this space being physically allocated (*definitions*) at runtime. A definition cannot be repeated twice inside the same block or a single identifier would refer to memory storage at two locations (which could then contain different values since C++ allows variables to be addressed through their addresses). A declaration that does not reserve memory at runtime, and therefore can be repeated arbitrarily many times, enables the compiler to process program constructs that employ the declared variable name in advance of a definition statement. For example, the type of a global variable **M** that is defined in an external program source file that will later be combined with a given file (through the linker) can be declared to the compiler through the declaration **extern int M;**. The compiler can then resolve subsequent statements that contain **M**.

The *scope* of a variable refers to the block (or possibly the file in a multifile program) in which it is defined. A variable with *global scope* exists in the global space outside all blocks (including the blocks enclosing function bodies) appearing in the program and is therefore accessible from within any block in the program (unless hidden by a local variable of the same name). For future reference, *namespace scope* is nested in the global scope through user-declared namespace definitions. *Class scope* applies to variables declared within a class body. As is evident from the above discussion of variable hiding, however, a variable is not necessarily *visible*, that is, accessible to program operations, within its scope.

5.8 rvalues and lvalues

An *lvalue* extends the concept of a variable by referring to any construct that accesses memory such that it can be assigned a value. Lvalues can therefore appear either on the left- or on the right-hand side of the assignment operator, =, which places the value of, or the value stored in, the construct to its right into the memory location associated with the lvalue to its left. Note that this is *not* the mathematical equality operator since mathematically contradictory statements such as **int m = 3; m = 2;** yield meaningful programming results. (To make this distinction clearer, many languages reserve a special symbol such as <— or a more abstract representation, :=, for assignment.) An *rvalue*, typified by a numerical constant, does not correspond to a program-accessible memory location and therefore must always appear on the right-hand side of the assignment operator; e.g. the statement **3 = m;** is clearly invalid.

5.9 Operators – precedence and associativity

An operator transforms or combines one or more expressions into new expressions. A unary operator acts on a single expression, such as the – sign in –5, whereas a binary operator resembles an implied function for which one argument is located on the left of the operator symbol and the second on the right.

Recall now that in the arithmetic expression $2 * 3 + 4$ multiplication is performed before addition unless parentheses are employed to indicate a different order of operations, as in $2 * (3 + 4)$, since the unary parentheses operator, which acts by evaluating the expression it encloses, is always applied before any arithmetic operators. This relative ordering of operations is termed *precedence*. Accordingly, the parenthesis operation possesses a higher level of precedence than the * and / operators, which in turn occupy a level above + and −. While 13 levels of precedence exist in C++, the basic structure is summarized by

```
-- Unary: (( ), ++, ::, new ...)              Highest
-- Arithmetic: (%, *, /, then +, −)
-- Bitwise shift/insertion, extraction: (<<, >>)
-- Relational: >, >=, <, <=
-- Logical: ==, !=, then && (and), then || (or)
-- Assignment: (=, += ...)                     Lowest
```

Often an additional rule is required to evaluate uniquely a statement containing several operators with the same precedence level. For example, in 5/4*3 the multiplication and division operators have equal precedence but (5/4)*3 and 5/(4*3) yield different results. However, in C++, *except where a meaningless result ensues, operators of equal precedence are left-associative*; that is, they are applied from left to right. Accordingly, 5/4*3 corresponds to (5/4)*3. Operators such as the assignment operator are right-associative since **m = n = 5** is then correctly interpreted as placing the value 5 first in the memory space associated with **n** and subsequently placing the value of **n** into the memory space of **m**. Left associativity would instead assign the value of **n** to **m** but then overwrite this value with 5.

Whenever knowledge of associativity or precedence is required in order to resolve an expression, unintentional errors frequently arise. Thus, in the above example, 5/(4*3) is very often intended but the parentheses are unintentionally omitted. Accordingly *parentheses should be employed wherever possible.*

5.10 The const **keyword**

The value of a variable that is defined by including the keyword **const** in its type specifier as in **const int** or **int const** cannot be changed by assigning a new value to the variable (although, as noted above, the variable can be hidden by a non-const variable of the same name in an inner block). Thus the second line below is illegal:

```
const int m = 3;
m = 4;                    // Compilation error
```

A **const** variable must be assigned a value when defined, otherwise it would possess a random value that could not subsequently be altered.

5.11 Formatting conventions

While the variable names and statement formatting in C++ can be chosen arbitrarily, subject only to the restrictions of Section 5.2, a logical set of conventions should be followed, such as

(1) Employ descriptive names. For non-object variables and functions, capitalize all words with the exception of the first, as in **numberOfGridPoints**.
(2) Place spaces to the right and left of binary operators, but not unary operators, e.g. **m = n + −1.0;**.
(3) Insert spaces after commas, the opening delimiters (and { and before the closing delimiters } and) as in **myFunction(int aI, int aJ);**, except for array indices.
(4) Indent each successive enclosed block by one further tab stop.
(5) Enclose segments of code that perform related functions with blank lines.
(6) Begin the names of function arguments with a small a.
(7) Capitalize names of classes, structures and objects.
(8) Begin names of internal class variables with a small i.
(9) Begin names of boolean variables with is and boolean functions with enable.
(10) Capitalize all letters of global constants.

5.12 Comments

A non-trivial program cannot be easily understood unless adequately commented and accompanied by a separate program description. To enable comments, the C++ compiler does not process text to the right of the delimiter // or between the delimiters /* and */. However, while the second procedure facilitates the removal of large blocks of code from compilation, if two non-adjacent segments of code are each enclosed in such delimiters and the line containing the first end delimiter */ is deleted by mistake, all code from the first occurrence of /* to the single remaining delimiter */ is ignored, yielding unexpected errors.

As a rule, a program should commence with "prolog" lines specifying a descriptive title, the revision number, author, revision date and program objective. Each section of code should be preceded by comment lines that explain its purpose. The interpretation of each significant statement should be placed either directly above or to the right of the line. Comments can contain data values that can be employed as test cases during debugging. To illustrate,

```
// comment.cpp
// Revision 1.0
```

```
// November 23, 2011
// Illustrates the use of comments
/* A C comment
extends through
several lines here. */
// This program demonstrates a procedure for avoiding division by zero

int main( int argc, char *argv[ ] )
{
    int testValue = 5;                  // Arbitrary non-zero value
// Uncomment this line to test division by zero
//   testValue = 0;
    if ( testValue != 0 ) cout << 10.0 / testValue;  // Safe function
}
```

Superfluous comments are, however, suppressed in the remainder of this text for space reasons.

Chapter 6
Procedural programming basics

The following three chapters introduce basic C++ program structure and syntax in the context first of procedural programming and subsequently, in the two later chapters, of object-oriented programming. The material in these chapters addresses the significant challenges encountered by beginning programmers.

6.1 Scientific software development

Procedural programing follows a clearly defined set of steps, which are discussed individually in this section.

Problem definition. First, a problem description that captures the main scenarios (possible outcomes), including possible abnormal situations, such as, for example, those generated by erroneous input data, should be formulated.

Detailed specification. A program specification comprises a detailed solution strategy such as the form and content of the input and output data, the equations to be programmed, the numerical methods to be employed, the hardware and software to be used and the manner in which the code will handle the various scenarios. This can be facilitated by first generating the input and output screens that the user will encounter.

Iterative coding and modular testing. Subsequently, the program tasks should be compartmentalized into functions. Each of these should be verified independently with a set of test data that is subsequently saved in comment lines for possible future use. Comments should be supplied for each additional function or block of code and the verified code modules packaged for reuse in other projects. As a rule, *only a single change or function is added to the program at a time before retesting.* In addition, *before implementing any non-trivial change, the previous version of the code (with an appropriate version number) should be saved in case an inadvertent error is introduced.* An editor that can identify the differences between two text files can then find subtle errors such as adding an additional (sometimes invisible) character during editing.

Equations in the program should match those in the specification document, while the variable names should be identical for both (assuming that the documentation will be carefully preserved) or the program should contain descriptive, compound names. In coding, *clarity should generally be favored over brevity and cleverness*, since far more time is generally required to develop and maintain a program than is required for program execution.

Initially, a program should solve only the simplest version of the problem and should ignore non-standard outcomes. After validation against a test case, typically an analytically tractable problem, the code can be incrementally enhanced to include additional scenarios and features. As new functionality is added, further requirements will surface while more efficient or transparent code structures become apparent. This favors an iterative programming methodology in which the original program specification is frequently updated as coding proceeds.

Final testing. Upon completion, the program should be verified against a set of test cases with known results, including those that branch into the structures which handle exceptional conditions. Test inputs are again preferably stored as comment lines within the program for future maintenance. If analytic results are unavailable, *a second program that solves the same problem with a different numerical or computational approach should be written*, since often the only reliable alternative is to verify every individual code line by inspecting each intermediate result with either a debugger or dedicated write statements.

Program maintenance. After a program has been completed, periodic updates and revisions are facilitated by employing high-level, properly commented modules that can be easily replaced, together with documents describing the program structure and operation.

6.2 The main( ) **function**

The entry point into (the first code executed by) a C++ program is the body of the **main( )** function. Accordingly, every C++ program must contain a single occurrence of

```
int main( int argc, char *argv[ ] )
{
... statements ...
}
```

or, more simply,

```
main( )
{
... statements ...
}
```

which can appear anywhere in the source file (program) after the external constructs appearing in **main( )** are declared. In the first of the two above implementations, **return 0;** or, equivalently in DEV-C++, **return EXIT_SUCCESS;** (the global constant **EXIT_SUCCESS** equals zero) should preferably immediately precede the closing brace.

6.3 Namespaces

C++ can group program elements into separate namespaces such that e.g. a function or variable in a namespace **A** is referred to from outside the namespace by appending a prefix **A::** before its name. The resulting ability to segregate code into non-interacting code segments facilitates library reuse. To illustrate:

```
namespace A {
   int M = 1;
}
namespace B {
   int M = 2;
}
using namespace std;

main( ) {
   int M = 3;
   cout << A::M << " " << B::M << " " << M << endl;   // Output : 1 2 3
}
```

If routines or libraries from numerous sources are employed in a program, this facility prevents the inadvertent inclusion of two similarly named program elements. However, when the programmer is certain that no such collisions will occur, the requirement that program elements in different namespaces be referred to with the corresponding prefix can be circumvented through the **using** directive:

```
namespace A {
   int M = 1;
}
namespace B {
   int N = 2;
}
using namespace A;
using namespace std;

main( )
{
   {
   using namespace B;
   cout << M << " " << N << " " << endl; // Output : 1 2
   }
}
```

The **using namespace B** is effective only inside its containing block, while the **A** and **std** namespaces are present from their **using** directives to the end of the program. If the **using namespace std;** statement is omitted in the above program, each element of the include files, such as **cout, cin, exp( )**, etc., must be individually prefixed with **std::** to indicate its membership in the **std** namespace. However, if **#include <iostream>** is replaced by the antiquated **#include <iostream.h>** directive, which does not enclose definitions in a namespace, **std::** must be omitted.

6.4   Preprocessor directives and #include statements

To assemble and transform code before compilation a primitive text editor, termed the *preprocessor*, is employed. Preprocessor commands, or *directives*, begin with the symbol # and are not terminated with a semicolon. This facility enables C++ to restrict the number of features available during compilation. Specialized routines for tasks such as mathematical operations and physical device access are then activated as required by incorporating appropriate "header" files through **#include** preprocessor directives followed by the name of the appropriate library package into a program before the features are employed. The **#include** statement reads a second file into the current file at the position at which the statement appears. It may adopt either of two forms, the first of which is

```
#include "includeFile"
```

which instructs the compiler to search for and incorporate a file named **include-File** first in the directory from which the program is being run and then, for most compilers, in the include-file subdirectory(s) of the compiler's main directory. Alternatively, writing

```
#include <includeFile>
```

leads to the compiler first attempting to locate **includeFile** in the compiler's include directories and then, in most cases, in other directories such as the user's directory. Some important include files are

```
#include <iostream>    // activates terminal and keyboard
                       // input and output
#include <fstream>     // activates input and output from
                       // the hard disk
#include <math>        // activates mathematics functions
                       // such as sqrt, sin
```

The preprocessor directive **#define A B** sets an expression, **A**, to a second expression, **B**, such that all instances of **A** are replaced by **B** before compilation. Hence, if **B** is a constant, **A** is replaced by this constant. Similarly, **#define square(x) x*x** replaces all instances of **square(m)** by **m * m** prior to compilation. In contrast, **A** in the statement **const typename A = B** represents a non-modifiable lvalue with an accessible storage address. Because textual substitutions circumvent type checking and scope resolution, **const** variables and **inline** functions (c.f. Section 6.25) are strongly preferred in C++. The directive **#define A** is also often employed together with **#ifdef** or **#ifndef**, **#else** and **#endif** directives to ensure that the same code lines are not compiled twice. That is, the block of code below starting with **int m**:

```
#ifndef MYFLAG
#define MYFLAG
```

```
int m;
... additional definitions and code statements
#endif
```

will be compiled only once even if it is included into a given program multiple times.

6.5 Arithmetic and logical operators

The arithmetic C++ operators include $+$, $-$, $*$ and $/$ together with the remainder operator %, which is defined such that **m % n** is the remainder when m is divided by n. *It is important to note that the result of this operation, unlike the modulus, is negative for negative numbers, which can give rise to unexpected errors.* Associated with each arithmetic operator is a compound assignment operator such that, for example, the statement **m += 2;** is identical to **m = m + 2;**. Additionally, the postfix and prefix operators, **m++**, **m−−** and **++m**, **−−m**, increment or decrement the value of **m** by unity so that **++m;** and **m++;** are equivalent. However, after

```
int n = 4;
m = n++;
```

$m = 4$, and $n = 5$, while

```
int n = 4;
m = ++n;
```

instead yields $m = 5$ and $n = 5$. That is, if $++$ *precedes* a variable name in a statement (the prefix operator), the $++$ operator is applied to increment the variable *before* the statement is evaluated, whereas if $++$ *follows* the variable name (the postfix operator), the $++$ operator is applied *after* evaluation of the statement. Statements containing both prefix and postfix operations yield different results for different compilers (since C++, unlike Java, does not specify the order of evaluation of such operations) and therefore must be avoided.

The C++ logical operations comprise **a == b** (equals), which yields 1 if the values of a and b are identical and 0 otherwise; **a != b** (not equals), which evaluates to 0 if a and b are identical and 1 otherwise; **a && b** (a and b), which evaluates to 1 only if both a and b are non-zero; and **a || b** (a or b), which equals 0 only if both a and b are 0. Finally, the unary operator $\sim$**a** (not a) yields 0 if a is non-zero and 1 otherwise. *Frequently, the second symbol is erroneously omitted* in == or **&&**, which invokes bitwise operators in place of the logical operators, as discussed in Section 14.14, typically resulting in a logical value of true. Addtionally, a space is not permitted between the two symbols in any compound operator such as <=, == or ++.

Finally, the lowest-precedence comma operator joins two expressions and returns the value of the expression on its right, so that the value of **k** in the expression **k = (m = 0, n = 1);** is 1.

6.6 The bool and enum types

Logical false in C++ is represented by any value with all zero bits, otherwise the logical value is true. Thus some representations of false are **int i = 0;**, **double d = 0.0;**, **char c = '\0';** (which has the numeric, ASCII, value 0 and is called the null character) and the intrinsically defined global constant **NULL**. These representations of false are equivalent when a variable can be legally assigned to a variable of another type so that **double d = 0.0; char c = d;** yields a zero character.

A **bool** type that accepts the values **false** and **true** is also often employed in logical statements. Assigning any non-zero value to a **bool** variable yields a value of 1. Sending (piping) the *manipulator* **boolalpha** to **cout** prints out the values of succeeding **bool** variables as either **false** or **true**; that is, **cout << boolalpha << false && true << endl;** yields the output **false** as opposed to 0 if **<< boolalpha** were omitted.

The **bool** construct is further generalized by the **enum** type. An **enum** type can be viewed as an integer type that additionally specifies a set of literal (i.e. named) constants called *enumerators* that can be assigned to variables of the type. Syntactically (a common convention is to capitalize **enum** variables), for

```
enum suites { HEARTS, SPADES, CLUBS = 4, DIAMONDS };
suites mySuite;
```

or, equivalently,

```
enum suites { HEARTS, SPADES, CLUBS = 4, DIAMONDS } mySuite;
```

mySuite can be set to any one of the four literal values **HEARTS, SPADES, CLUBS** or **DIAMONDS**. The values **HEARTS, SPADES, CLUBS** and **DIAMONDS** are then numerically equivalent to 0, 1, 4 and 5, respectively, as evidenced when a **suites** variable containing one of the literal values is converted to an **int** as below:

```
enum suites { HEARTS, SPADES, CLUBS = 4, DIAMONDS };
suites firstSuite = 3;
suites secondSuite = DIAMONDS;
cout << firstSuite << '\t' << secondSuite << endl;
// Output : 3    5
if ( secondSuite == DIAMONDS ) cout << "Diamond" << endl;
// Output: Diamond
```

6.7 Control flow, if statements and implicit blocks

Control constructs direct program flow according to the outcome of logical operations. Viewing blocks as the equivalent of paragraphs, control statements determine whether and in what order these paragraphs are executed at runtime. For example, the **if** statement

```
if ( A ) {
... statements ...
}
```

executes the block labeled **statements** when the logical expression **A** evaluates to **true**.

Control constructs share several features that lead to programming errors. If the block governed by any control construct contains a single statement, the enclosing braces can be omitted, but the block structure is still implicitly present.

```
if ( A ) statement;
```

Although the above compound statement is often written on two lines, this should be avoided where space permits, in order to avoid the following errors. First, separating a control condition from the following statement by a semicolon,

```
if ( A ); statement;      // WRONG!
```

mistakenly places a null statement under the control of **A** so that **statement** is always executed. Next, in

```
if ( A ) int m = 1;
```

although braces do not surround **int i = m;** the statement is still implicitly enclosed in a separate block. Therefore **m** is subsequently destroyed and is unavailable to the remainder of the program. Conversely, *braces are often mistakenly omitted in multi-line if statements*. Only the first of the statements following the **if** statement is then influenced by the control condition. Finally, *troublesome errors occur when the assignment operator = is employed in place of the logical equality operator == as in the statement* **if (B = C)**. As a result, the value of **B** is initially set to that of **C** and the logical statement is then evaluated with this unintended value as an argument. Typically, **C** differs from zero, so that the logical statement evaluates to **true** and the subsequent block is executed.

The extent of an **if** statement can be extended by appending an **else** statement that is executed if the assertion in the **if** statement is false. An abbreviated form for **if (A) {B} else {C}** is the ternary conditional operator **A ? {B} : {C}**. If the **else** keyword is followed by an **if** statement, **else** and **if** are generally placed on a single line, as in **if (!grade) {...} else if (grade == 1) {...} else { }**.

Repeated **else if** statements can be collected into a single **switch** construct. Because of the subtleties of the syntax, the following example should be consulted during coding:

```
int grade = 8;
switch ( grade / 10 ) {
        case ( 7 ): cout << "C" << endl; break;
        case ( 8 ): cout << "B" << endl; break;
        case ( 9 ): cout << "A" << endl; break;
        default: cout << "F" << endl;
}
```

The **break** statements in the above program transfer control to the statement immediately following the **switch** block. Removing these statements leaves control within the block so that the (optional) **default:** statement is always executed. A **switch** construct can be applied to other types of variables, as in **char c = 'b'; switch(c) { case ('a'): ... }.**

6.8 The for **statement**

A **for** statement executes the block under its control while incrementing one or more variables until a certain logical condition is fulfilled according to the following syntax:

```
for ( initialization (I); termination (T); statements (S) )
{ body (B);}
```

Here I, T and S represent any number (including zero) of statements separated by commas and B is a code segment. The statements are executed in the order: I, T, B, S, T, B, S, ..., T.

The **for** statement is normally encountered with the following format:

```
for ( int loop = 0; loop < 5; loop++ ) cout << loop << ' ';
// output: 0 1 2 3 4
```

where **loop++** and **++loop** can be interchanged. Again, since employing **i** or **j** as loop-variable names invites collisions with similarly named variables elsewhere in the program, identifiers such as **loop, loopInner** and **loopOuter** are highly recommended. Except in older compilers, a loop variable that is defined in the initialization statement is considered to be defined inside the body of the loop and is destroyed when the **for** block is exited for any reason. Thus, to access the value of **loop** after the **for** block terminates, it must be defined outside the loop body, as in

```
int loop;
for ( loop = 5; loop > 0; loop-- ) cout << loop << ' ';
// output: 4 3 2 1
cout << endl << loop << endl;
// output: 1
```

Omitting the condition statement yields an infinite (non-terminating) **for** loop, as in **for (loop = 0 ; ; loop++) { ... }** or **for (; ;) { ... }.** The statements enclosed within the block then should poll for a certain user or system response to exit the loop.

6.9 while **and** do ... while **statements**

The first **for** loop of the previous section can be recast through a **while** statement as

```
int loop = 0;
while ( loop < 5 ) {
      cout << loop++ << "  ";
}
```

or alternatively with a **do ... while** statement as

```
int loop = 0;
do {
      cout << loop++ << "  ";
} while ( loop < 5 );
```

Note that a terminating semicolon is required for a **do ... while** statement but not for the **while** statement. However, if **int loop = 6;** is substituted for **int loop = 0;** above, only the **while** statement reproduces the results of the **for** loop since the logical condition in the **do ... while** construct is evaluated after rather than before the statement block is executed.

6.10   The break, continue **and** exit( ) **statements**

A **break** statement, which can be placed only within a control structure, transfers program control to the first statement following the end of the (innermost) structure containing the **break**. In contrast, a **continue** statement in an iteration ends the current iteration, after which the termination condition, T, is evaluated. If false, the subsequent iteration commences, as in

```
for ( int loop = 0; loop < 5; loop ++) {
      if ( loop == 1 ) continue;
      if ( loop == 3 ) break;
      cout << loop << ' ';                      // output: 0 2
}
```

The break statement can be employed to exit manually from a control construct. That is, placing

```
int r;
cout << "Enter 0 to terminate ";
cin >> r;
if ( r == 0 ) break;
```

inside a control loop enables termination of the loop from the keyboard; the variable **r** is then termed a *sentinel*. An alternative procedure is

```
int r = 1;
while ( r != 0 ) {
      statements;
      cout << "Enter 0 to terminate ";
      cin >> r;
}
```

A running program can be terminated with the **exit(int aM)** function (here **#include <stdlib.h>** may be required). While the function parameter, which is normally set to EXIT_SUCCESS or equivalently zero or EXIT_FAILURE or any non-zero value for normal and abnormal program termination, respectively, must be supplied, this does not affect the program and is largely invisible to the user.

6.11 The typedef **keyword**

A **typedef** statement enables a new identifier to be assigned to a particular type. The type of all variables within the scope of the **typedef** can then simultaneously be altered by replacing the identifier and recompiling. For example, to change all **double** variables to **float** in the program

```
typedef double myType;

main( ) {
        myType d1 = 3.0;
        myType d2 = 4.0;
        cout << d1 << '\t' << d2 << endl;
}
```

the **typedef** statement should be replaced by **typedef float myType;**

6.12 Input and output streams

Following the occurrence of the **#include <iostream>** statement, the so-called standard input and output streams **cin** and **cout** (pronounced see-in and see-out) are enabled. These streams implement intelligent buffers between the program and the standard input and output devices. The **cin** stream accepts character or numeric input from the keyboard that is extracted (piped) from the stream and placed into a program variable through the extraction operator >>. Accordingly,

```
int m, n;
cin >> m >> n;
```

reads first the value of **m** and then that of **n** from the input stream. Since C++ equates single and multiple whitespace characters, two values entered from the keyboard separated by any combination of whitespace characters such as spaces, tabs or carriage returns (but not commas!) are stored successively in **m** and **n**.

Similarly, values are piped from the program into the standard output stream **cout** that is attached to the terminal through the insertion operator <<. For example,

```
cout << m << '\t' << n << '\n' << flush;
```

or, equivalently,

```
cout << m << '\t' << n << endl;
```

displays the value of **m**, a tab ('**\t**' is called a tab character) and the value of **n** and terminates with a newline character '**\n**'. The **endl** statement corresponds to '**\n**' followed by **flush**. The keyword **flush** insures that the contents of the stream buffer are displayed on the terminal, while a **cout** statement that does not end with **flush** or **endl** can delay displaying the output. In this case, a program that terminates abnormally after the output statement, but before the output appears on the terminal, will incorrectly imply that the error occurred before the statement. Other formatting and nonprinting characters include

'**\v**'   \\ vertical tab
'**\b**'   \\ backspace
'**\r**'   \\ carriage return
'**\f**'   \\ formfeed
'**\a**'   \\ alert (bell)
'**\?**'   \\ question mark
'**\"**'   \\ single quote
'**\"**'   \\ double quote
'****'   \\ the \ character.

6.13   File streams

Just as the input and output streams **cin** and **cout** interface memory buffers to the keyboard and terminal, streams that similarly interface physical storage locations are enabled by adding the line #**include <fstream>** to the beginning of the program (in many compilers the **fstream** header file contains **iostream**). However, unlike the unique "standard" keyboard and terminal, to access a file, its name must first be specified as in

```
fstream fs( "input.dat" );
```

Here the disk file **input.dat** in the directory from which the program executes is associated with a stream, **fs**, that can be employed for both input and output. To limit the stream to either input or output (corresponding to **cin** or **cout**) **fstream** should be replaced by **ifstream** or **ofstream**, respectively. The input file **input.dat** can be created by entering the exact keystrokes including whitespace characters into a text or program as would be entered into the program from the keyboard through **cin**.

6.14   Casts

In C++, conversion of built-in variables between closely related types occurs automatically, as in **char c = 56;** in which the integer 56 is converted to the character '**8**' (the 56th member of the ASCII character set). If variables of different types that can be automatically converted into each other are combined through a binary operator as in

```
char c = 'a';
int m = 10;
cout << ( c + m ) << endl;        // Output: 107
```

then the variable that occupies less memory space (in this case the **char**) is converted (promoted) to the type of the other variable (an **int**) before the operator is applied.

Conversions can be forced through explicit *type-conversion operators* (casts) as in **cout << char(m) << endl;**. Casts have several additional forms in C++. A *C-cast* is written **(char) m**. The syntax **static_cast<char>(m)** forces the conversion of **m** from an **int** to a **char** at compile time. Casts with system-dependent behavior and those that remove the **const** property of variables are segregated into calls to **reinterpret_cast<typename>** and **const_cast<typename>**. Finally, **dynamic_cast<typename>** reserves the implementation of the cast until runtime.

6.15 Functions

A C++ function acts on a set of input variables and returns zero or one output variables. Variables defined within a function are normally isolated from the remainder of the program, facilitating testing, implementation and subsequent reuse by other programs. A modular program consists almost exclusively of functions and control structures, thus representing a physical problem as a logical flow among individual program elements. Code for a function should thus perform a single, well-defined task.

A general function can be represented as **returnType myFunction(type1 aP1, type2 aP2, ... , typeN aPN)**. The input variables **aP1 . . . aPN** are termed *arguments*. The convention that all argument names begin with a lower-case **a** greatly enhances programming clarity. The function is called (activated) through a statement such as **returnType x = myFunction(p1, p2, . . . , pN)**; in which **p1, . . . , pN**, which can in general be either constants or variables, are referred to as *formal parameters* or more simply as parameters. A function can possess any number (including zero) of arguments but can only return zero or one values. *The return type of a function that lacks a return value must be specified* as **void**, except for constructor (and destructor) functions as noted in Section 8.5. (In many situations C++ assumes a default type of **int** when no type is specified; for example, **const i;** is generally interpreted as **const int i;** while **f( ) { return 5; }** is compiled as **int f( ) { return 5; }** and is not a **void** function.) A function without arguments must still be called with parentheses as in **int j = f( );**.

6.16 Principles of function operation

Depending on the outcome of logical conditions, a function in a C++ program can be invoked any number, including zero, of times during program execution.

Since a function can contain numerous commands and allocate arbitrary amounts of memory for its internal variables, memory space for the function is not automatically assigned when a program is initialized, unlike space for global variables or variables defined within **main( )**. Rather, the machine-language instructions associated with the function body are written to memory. The function name in the calling program is then associated with the 4-byte (for a 32-bit machine) memory address of the start of this instruction set or *function record*, as is verified by

```
void print( ) { cout << "test"; }

main ( ) {
    cout << print << endl;   // Note: print is not followed by ( )
}
```

This displays a hexadecimal number such as 0040115E on a 32-bit compiler as memory addresses are by default expressed by C++ in hexadecimal format, for which the numbers 0–15 are represented by the numbers 1–9 followed by A–F. Each group of two hexadecimal numbers together then corresponds to 8 bits or 1 byte; hence the eight hexidecimal numbers above represent 4 bytes or 32 bits corresponding to the address of a 32-bit storage location. Note that **print** in **main( )** does not contain parentheses, which would instead evaluate (here activate) the instructions at this location – since the effect of the parenthesis operator is to evaluate the expression that it encloses.

When a function is called, the operating system reads the stored machine-language instructions from the starting memory location onward. These commands first reserve space in memory for the argument variables and then copy the values of the function's formal parameters into this newly allocated memory. Therefore, the sequence of statements executed when a function such as **void f(int aM) { int b = 2; ...}** is called and its argument variables defined and initialized can be represented by **void f(%1) { int aM = %1; int b = 2; ... }**, where **%1** denotes the parameter value obtained from the calling routine. The statements in the function block are executed until the end of the block (for a **void** function), when control is passed back to the calling program, or a **return** statement is encountered that additionally passes the return value. The operating system then deallocates the memory occupied by the function and its variables, returning to the initial state in which the function name refers to the corresponding stored binary instruction set.

Since a function invocation creates new, independent memory locations for the function arguments, and then copies the values of the function parameters into these locations, *altering the function arguments inside the function body* as for **aM** in the example below *leaves the parameter* **m** *inside* **main( )** *unaffected*, a feature termed *call-by-value*:

```
void change( int aM ) { aM = 2 * aM; }

main ( ) {
```

```
        int m = 3;
        change ( m );
        cout << m << endl;                    // output 3
}
```

6.17 Function declarations and prototypes

A statement such as **void change(int aM);** or equivalently **void change(int);** (the semicolon is required) is termed the function *prototype or signature*. While the prototype supplies sufficient information for the compiler to typecheck statements involving **change**, memory is not allocated, since the size of a function in memory corresponds to that required to store the binary representations of the commands appearing in its body (i.e. the function block). Therefore, the prototype constitutes a function declaration. Once supplied, a single function definition, consisting of the prototype followed by the function body, can be located anywhere in the program or placed in a separate program that is later compiled or linked together with the code. Thus, the previous program can be rewritten as (this form is actually required in C)

```
void change( int );

main ( ) {
        int m = 3;
        change ( m );
        cout << m << endl;                    // output 3
}

void change( int aM ) { aM = 2 * aM; }
```

6.18 Enumerators and functions

By employing enumerators as function arguments or return values the function parameters and return values can be restricted to meaningful quantities. The following code, for example, determines whether a voltage is sufficiently large enough to activate a connected device:

```
enum isEnabled { off, on };
isEnabled lightSource ( double aVolts ) {
        if ( aVolts > 0.2 ) return on;
        else return off;
}
```

6.19 Overloading and argument conversion

Since C++ typechecking resolves both the number and the type of function arguments, a function is uniquely specified by this information together with its name. The compiler in fact *mangles* (joins) the function name with those of its arguments to generate a new function identifier. Thus, **int change(int);** and **int**

change(double); become separate entities distinguished by different mangled names. The expression **change(2)** invokes the first of these, while **change(2.0)** calls the second. This function *overloading* represents a form of *polymorphism*, which means the behavior of a function is determined by its environment here manifested by the type and number of arguments. However, a function such as **double change(int);** with the same name and argument list as an existing prototype but with a different return type cannot be additionally defined since C++ cannot resolve a function by the type of the variable assigned to the return value.

The expression **void f(int aI)** with e.g. **double pI = 2.5; f(pI);** implicitly executes the statement **int aI = pI;**, so that **pI** is automatically transformed in the function body to 2. Therefore, automatic type conversions occur between a function parameter in the calling program and the corresponding function argument. However, if multiple versions of an overloaded function possess the same number of arguments, as in the two **change** functions above, a rule must be present to determine which, if any, will be utilized if the function is called with an argument of a third type such as, for example, in **char c = 'a'; change(c);**. While a full discussion of this topic falls beyond the scope of this text, simple logical considerations generally apply. Here, since the size of a **char** is closer to that of an **int** than to that of a **double**, the **int change(int);** version is employed.

6.20 Built-in functions and header files

The C++ compiler and its include files provide extensive libraries of functions and predefined constants. However, knowledge of their argument and return types is essential in order to avoid errors. For example, in many C++ compilers the built-in absolute-value function possesses a prototype

```
int abs( int );
```

Errors can therefore arise if this function is called with a double argument expecting that a **double** rather than an **int** will be returned. Unless the absolute value of the difference between **a** and **b** exceeds 1 below, the **abs** function will in such cases return 0, and the logical condition in the **if** statement argument then evaluates to true:

```
double a, b, eps = 1.e-5;
cin >> a >> b;
if ( abs( a - b ) < eps ) { ... statements ... }
```

The desired function for the above program is

```
double fabs( double );
```

in place of **abs**, which must be preceded by the preprocessor directive

```
#include <math.h>
```

The math library additionally contains a number of important mathematical functions that operate on double arguments. These include **pow(x, n)**, which raises **x** to the (integer or non-integer) power **n**, **sqrt(x)**, **fabs(x)**, **ceil(x)**, which rounds x upward, **floor(x)**, which rounds x downward, **exp(x)**, **log(x)**, **log10(x)**, **sin(x)**, **cos(x)**, **tan(x)**, **tanh(x)**, **asin(x)**, the arcsin function, and other inverse functions. The math.h library also defines global constants such as pi (normally labeled **M_PI**).

To determine the full set of functions and constants present in an include file for a given version of a compiler, the header file should be examined directly. To locate the math.h header file, navigate to the X:\Dev-Cpp\include directory through e.g. the My Computer icon on your desktop, where X: is the installation drive letter. Double clicking on the icon for math.h opens the notepad editor. Browsing through the file reveals the lines

```
_CRTIMP double __cdecl sin (double);
_CRTIMP double __cdecl sinh (double);
_CRTIMP double __cdecl atan2 (double, double);
...
```

as well as the following partial list of predefined constants:

```
#define M_E                     2.7182818284590452354
#define M_PI                    3.14159265358979323846
#define M_PI_2                  1.57079632679489661923
...
```

Professional programmers frequently employ this technique to extract comprehensive, current information on the language implementation.

6.21 The assert **statement and try and catch blocks**

Design by contract is a software-engineering technique in which the programmer specifies for each function the physically valid ranges for the input and output values. These preconditions and postconditions (the "contract") are checked upon entering and exiting a function through appropriate logical statements, which are often introduced as arguments to the **assert( )** function, which terminates program execution if its argument evaluates to logical false as in the following example:

```
#include <assert.h>

main ( ) {
     int m = 1;
     assert ( m == 2 );              // test for an error condition
}
```

which generates the output

```
Assertion failed: m == 2, file assert.cpp, line 5
```

For production runs, the preprocessor directive **#define NDEBUG** precludes the compiler from processing all subsequent assert statements.

Preconditions and postconditions can also be implemented through **try** and **catch** blocks. When a **throw** statement is invoked within a **try** block, program control passes to the **catch** block with the closest valid match to the arguments of the **try** block. After the **catch** block has terminated, the program resumes unless an **exit** statement is encountered inside the **catch** block. A variable defined within a **try** block is destroyed when the block is exited. To insure in this manner that the input to a function **checkSquare(double aT)** is positive while its output is ≤100:

```
#include <stdio.h>                        // includes the exit( )
function.
double checkSquare ( double aT ) {
      // precondition: The input argument is > 0.
      double result;
      try {
            if ( aT < 0 ) throw "Input must be larger than zero";
            double mySquare = aT * aT;
            // postcondition: The result is <= 100.
            if ( mySquare > 100 ) throw result;
      }
      catch ( double aResult ) {
            cout << "Invalid result = " << result << endl;
            exit ( 0 );
      }
      catch ( char aMessage[80] ) {
            cout << aMessage << endl;
            exit( 0 );
      }
      return result;
}
```

6.22 Multiple return statements

While a function in C++ can return only a single variable, it can possess arbitrarily many return statements, each of which exits the function block. For example, **get( )** below returns the character 'y' when 1 is entered from the keyboard and 'n' otherwise:

```
char get ( ) {
      int m;
      cout << "Insert a value" << endl;
      cin >> m;
      if ( m == 1 ) return 'y';
      return 'n';
}
```

6.23 Default parameters

Sometimes a function argument differs from a standard value only in unusual circumstances. The argument can then be assigned default parameter values that are changed only if a different value is explicitly specified in the function call. *Default parameters must appear at the end of the function prototype* (otherwise e.g. a call **f(3, 4)** to a function **f(int a = 2, int b, int c = 3)** would be ambiguous with respect to which parameter is overwritten). Further, *default parameter values can be specified once only*, either in one and only one declaration statement or in the function definition. Hence,

```
void myFunction( int a, int b );
void myFunction( int a, int b = 1 );

main ( ) {
    myFunction( 3 );                    // output: 3   1
    myFunction( 3, 2 );                 // output: 3   2
}

void myFunction( int a, int b ) { cout << a << '\t' << b << endl; }
```

constitutes a valid program, but the default parameter value cannot be additionally introduced either into the first function declaration or into the function definition.

6.24 Functions and global variables

While a function C++ can return only zero or one variable, it can modify any number of *global variables* that are defined in the global scope outside all function blocks. Such variables can be accessed by the bodies of all subsequently defined functions unless hidden by a second variable with the same name. Thus the **change( )** function below modifies two global variables, which are subsequently displayed in **main( )**:

```
int GLOBALI, GLOBALJ;
void change( ) { GLOBALI = 2; GLOBALJ = 3; }

main( ) {
    GLOBALI = 0;
    GLOBALJ = 1;
    change( );
    cout << GLOBALI << '\t' << GLOBALJ << endl;   // Output: 2   3
}
```

Such a technique, however, can lead to severe errors, since an inadvertent change to a global variable in any section of the program propagates to all other program units. This can yield unforeseen side effects separated by a large code distance from the error source, impeding its identification.

6.25 Inline functions

The body of a function that is declared inline is automatically substituted into each function call before compilation. Hence for

```
inline int square( int x ) { return x * x; }
```

a call to **square(4 * y)** is replaced by **(4 * y) * (4 * y)**. The code then executes more rapidly, since a function record does not have to be processed upon each call to square, but at the cost of increased compile times and object file sizes. Generally, **inline** declarations should be reserved for small functions (however, functions defined as opposed to declared within a class body are automatically inlined). An inline function does not possess a prototype, since the body of the function must be accessible to the compiler each time the function is called to enable the required substitutions.

6.26 Recursive functions

Since in a function definition the body is preceded by the prototype, the compiler can already typecheck calls to the function itself from within its body. Such a function is then termed *recursive*. As an example, the factorial function, defined as $n! = n(n-1)!$, with $1! = 1$, can be written

```
double factorial ( const int aN ) {
      double temp;               // a floating-point type is required
                                 // for large aN
      if ( aN == 1 ) temp = 1;
      else temp = aN * factorial ( aN - 1 );
      return temp;
}
```

Thus, for **int k = factorial(3);**, in the first call to **factorial(3)**, **temp** is set to **3 * factorial(2)**, which cannot be evaluated until **factorial(2)** returns. However, when **factorial(2)** executes, a new **temp** variable in a separate memory space is created for **2 * factorial(1)**. Finally **factorial(1)** returns 1, enabling **factorial(2)** and then **factorial(3)** to complete. While the overhead associated with multiple function calls and the simultaneous allocation of memory resources can be considerable, this is often outweighed by increased programming simplicity.

As another example, the square root of 2 can be expressed by the continued fraction

$$\sqrt{2} - 1 = \frac{2-1}{\sqrt{2}+1} = \frac{1}{1+\sqrt{2}} = \frac{1}{2+(\sqrt{2}-1)} \approx \cfrac{1}{2 + \left(\cfrac{1}{2 + \cfrac{1}{2+\cdots}} \right)}$$

A numerical algorithm is obtained after placing a zero at the location of the ellipsis ($\ldots$). A program that computes $\sqrt{2}$ for one to ten retained terms in the continued fraction is

```
main ( ) {
    int maxLoop = 20;
    for ( int outerLoop = 1; outerLoop < maxLoop; outerLoop++ ){
        double root = 0;
        for ( int loop = 0; loop < outLoop; loop++ ){
            root = 1. / ( 2. + root );
        }
        cout << 1. + root << endl;
    }
}
```

6.27 Modular programming

A modular program is principally formed from declarations and definitions, control logic and function calls, as in

```
int getInput ( ) {
    int r;
    cin >> r;
    return r;
}
int multiply( int a1, int a2 ) {
    return a1* a2;
}
void print( int aI ) {
    cout << aI << endl;
}

main ( ) {
    int input1 = getInput ( );
    int input2 = getInput ( );
    int result = multiply( input1, input2 );
    print ( result );
}
```

Such code is easily analyzed and corrected, since each function can be tested separately.

6.28 Arrays

An array comprises an indexed set of variables of the same type. The array definition **int v[3];** reserves memory for an array of three **int** elements and associates the type of **v** with that of an array of **int** elements. The content of the storage location for the ith element is accessed by following the array name with the index operator **[ ]**, as in

```
main ( ) {
      int v[3];
      for ( int loop = 0; loop < 3; loop++ ) v[loop] = loop;
      cout << v[1] << endl;                    // Output: 1
}
```

The syntax

```
int v[3] = {1, 2};
```

which can be employed only when the array is defined, initializes the first element of **v** to 1, the second element to 2 and all remaining elements to 0.

6.29 Program errors

Coding a program invariably leads to numerous errors, most of which can be identified by analyzing the resulting, often cryptic, computer-generated messages. Beginning programmers should therefore carefully examine these, which fall into three categories.

Compiler errors. Most frequently, the compiler cannot interpret a program because of incorrect syntax or program structure. Some examples are **m = n / * l;**, writing **m = n;** without first defining **m** through the statement **int m;**, omitting the semicolon at the end of a line so that the line is joined with the subsequent line and confusing characters such as 1 (one) and l (the letter l) and O (the letter O) and 0 (zero). For each such instance an error message with a corresponding line number is generated, which, however, unfortunately, typically is unrelated to the source of the problem. For example, if a variable is not declared, the compiler cannot process any statement in which the variable is present. Therefore, numerous error messages with line numbers remote from the error source result. Accordingly, compiler errors are best corrected by resolving the first few errors in the error list and recompiling. All messages resulting from dependences on the incorrect lines vanish and the greatly reduced number of remaining errors can be similarly eliminated in further stages.

As an example of a compile error (in command-line Borland C++), the program

```
main ( ) {
      m = 2;
      cout << m << endl;
}
```

yields the following error message in DEV-C++:

```
2 C:\programs\test.cpp 'm' undeclared (first use this function)
```

where the .cpp extension indicates that the error originates from a source-code file.

Link errors. Typically, link errors result when a program attempts to call a function or use a variable that is not defined either in the program file or in the additional files that are linked with the program, as in the program below:

```
int mySquare( int );   // declares but does not define the function
main ( ) {

    int m = 3;
    cout << mySquare( m ) << endl;
}
```

which yields the message

```
[Linker error] undefined reference to 'mySquare(int)'
```

The missing function definition (e.g. function body) required by the linker must accordingly be supplied in the source code or in a second object file.

Runtime errors. The most severe mistakes arise during program execution. For example, a program can fail because of an incorrect numerical algorithm or improper program flow. These issues are often most easily detected by comparing the program results with those of an analytic evaluation or of a second program that employs a different calculational approach. Alternatively, a construct can be syntactically correct but yield unintended results; for example, if **cin >> m, n;** is employed in place of **cin >> m >> n;**, only the first input value is read into **m**. An incorrect result or overflow or underflow condition could then be generated with or without an accompanying error message.

Even more subtle runtime errors are associated with improper memory access. If a program reads or writes to a memory address beyond the space allocated to the program through, for example, an array index equal to or larger than the number of elements allocated to the array, the operating system often intercepts the illegal access. In this case, the program is terminated and a system error message such as "segmentation fault" is displayed inside a pop-up window (i.e. the program is addressing memory outside the segment permitted by the operating system). However, if the array variable accesses a memory location of another variable in the program, seemingly random errors occur when the program is executed with different input values.

6.30 Numerical errors with floating-point types

Calculations performed with floating data types are further subject to *round-off errors*. While each arithmetic operation maintains a high level of accuracy, combinations of operations can substantially degrade precision under certain conditions. This effect can be quantified in terms of the *machine epsilon* or machine accuracy, which is the smallest floating point number, ε_m, such that $1.0 + \varepsilon_m \neq 1.0$. The machine-epsilon values for float and double precision numbers are given by the constants FLT_EPSILON and DBL_EPSILON defined in the <**float.h**> header file. In contrast, the overflow and underflow thresholds correspond to the largest and smallest number that each data type can represent, namely $10^{\pm 38}$ for a **float** (FLT_MIN and FLT_MAX) and $10^{\pm 308}$ for a **double** (DBL_MIN and DBL_MAX). In this context, it should be noted that a number that exceeds the overflow bound or is obtained by dividing zero by itself is represented by a symbolic value such as **Inf** or **NaN**, which subsequently propagates through the calculation according to rules such as **1 / Inf = 0** and **2 * NaN = NaN**.

While the roundoff error of a single calculation is approximated by ε_m, the signs of these terms fluctuate if the calculation is repeated N times. Accordingly, the combined error describes a random walk as a function of N, yielding on average a total error of $N^{1/2}\varepsilon_m$. However, the *relative* error in the difference of two nearly equal quantities, such as a function computed at two closely separated points, with different roundoff errors, $C(1 + \delta + \varepsilon_m)$ and $C(1 + \varepsilon'_m)$, where $\varepsilon \ll \delta \ll 1$, is $C(\varepsilon' - \varepsilon)/(C\delta) \approx \sqrt{2}\varepsilon/\delta \gg \varepsilon$. *Accordingly the number of significant digits in the result that propagates to subsequent calculations is reduced* by $\approx -\log_{10}\delta$. Such cancellations, which occur, for example, in the numerical derivatives of nearly constant functions and in the quadratic formula for $b \approx \sqrt{b^2 - 4ac}$, yield significant errors and must be carefully avoided.

A more serious issue relates to the stability of a numerical algorithm. Consider the computation of the path of a ball starting at $x = y = z = 0$ and rolling in the z-direction along the line given by the maximum of the parabolic surface defined by $y = -x^2$ for all values of z. Any numerical error increases with time, leading to rapidly increasing or decreasing values of x. Numerical methods that respond in a similar fashion to rounding errors or numerical fluctuations are termed *unstable* and yield intrinsically unreliable results unless the accumulated effect of the error terms is carefully analyzed and limited.

Chapter 7
An introduction to object-oriented analysis

The structure of object-oriented programming, which differs significantly from that of procedural programming, must be thoroughly understood before attempting code development. The construction of an object-oriented framework for a physical problem is termed object-oriented analysis, while the translation of the framework into actual C++ code is the domain of object-oriented design. These topics are discussed separately in this and the subsequent chapter.

7.1 Procedural versus object-oriented programming

Consider calculating and graphing the trajectory of one or more vibrating springs with attached masses. A procedural approach inserts data describing the physical system into a series of functions, each of which transforms input information into output values. Control statements sequence the functions according to the outcome of certain logical conditions. Schematically, for a single spring

```
main ( ) {
    float position[50], velocity[50], springConstant, mass,
        timeIncrement;
    int numberOfSteps;
    cin >> position[0] >> velocity[0] >> springConstant
        >> mass >> numberOfSteps >> timeIncrement;
    propagate( position, velocity, timeIncrement, numberOfSteps,
        mass, springConstant );
    plot( position, velocity, numberOfSteps );
}
```

where the initial position and velocity are stored in the first elements of the 50-component arrays **position[50]** and **velocity[50]**. The **propagate( )** routine calculates subsequent positions and velocities by advancing time over **numberOfSteps** time steps of duration **timeIncrement**. Finally, the arrays are graphed by the **plot( )** routine.

While procedural programming generally proves optimal for small programming projects, difficulties are encountered when applied to complex systems. First, a program that takes into account all physically realizable situations generally requires a description of the interaction of a physical object with its environment. For example, in the above program the spring could be excited by a launching device and its motion recorded by an appropriate measurement system. Each of these objects in turn has adjustable settings and is influenced by interactions with additional system components. While additional functions can be introduced, their relationships become increasingly involved. Further, if several springs with different loads and spring constants are present, numerous sets of related variables must be present, either as separate one-dimensional arrays and variables **position1[50]**, **velocity1[50]**, **mass1**, **position2[50]**, ... or as single-dimensional and multidimensional arrays, **position[40][50]**, **mass[40]**, etc. If several types of springs exist, with differing behaviors such as spring constants that age differently, a particular function must additionally be associated with each spring. Object-oriented programming addresses the issues above through the techniques of encapsulation, polymorphism and inheritance.

Encapsulation and *information hiding* are implemented through objects and classes. In a procedural program, the highest-level self-contained program unit is a function, which performs a specific task on a set of input data through a series of related instructions. However, a physical *object* in the real world exhibits an integrated structure. To illustrate, a car possesses a set of *properties*, corresponding to *internal data* or *member variables*, such as the amount of stored gas and battery charge, the currents in the electric circuits, the speed of the mechanical subsystems and so on. Further, the car realizes certain *behaviors* according to its principles of operation, typified by the relationship between its acceleration and the angle of the accelerator and brake pedals. These can be associated with *methods* or *member functions*. A description of the car should therefore form a self-contained program unit that describes both the properties and the behaviors of the physical device. Of course, the complete set of these is unmanageably large, since it could include a description of every part or even atom in the vehicle. However, only a few high-level features are of practical relevance. Taken together, these yield a meaningful *abstraction* of the object. In object-oriented programming, the representation of this abstraction as a self-contained set of internal variables and internal functions is termed *encapsulation*.

Returning to the car example, numerous significant properties and behaviors such as the voltages in the internal circuits and the angular velocities of various components are inaccessible to the user, who interacts with the car through a highly simplified interface consisting of the steering wheel, pedal, etc. In object-oriented terminology, *public* internal variables or functions of an object are those that can be accessed by any external user, while *private* variables are available only to components of the object itself. In a programming context, public variables or functions can be examined from anywhere in the program, while *private* variables can be accessed or changed only by the internal functions of an object. This *information hiding* segregates internal object properties from the remainder of the program and therefore from undesired external interactions.

Polymorphism refers to assigning context-specific behaviors to a single program entity such that, for example, the behavior of a function is determined by the context of the object within which it resides. For example, the evolution of the spring constant with repeated use would be different for a **LoadedMetalSpring** object than for a **LoadedPlasticSpring** object. In the same manner, the effect of double clicking on a document icon on a computer screen depends on the document type.

Finally, object-oriented programming incorporates the additional facility of *inheritance*. This enables variables and functions associated with one class of objects, such as **LoadedSpring** objects, to be automatically incorporated into a second class, such as **LoadedParallelSprings**, consisting of two or more joined springs, without copying shared code. Instead, **LoadedParallelSprings** objects retain all features of **LoadedSpring** objects, except those that are explicitly redefined or added, through a single keyword indicating that **LoadedParallelSprings** inherits from **LoadedSpring**.

7.2 Problem definition

In principle, an object-oriented program is developed by identifying the interacting objects (object discovery) and their relevant interactions. However, since numerous objects generally affect the system behavior, a trade-off arises when defining the boundary between the system and its exterior. Accordingly, the problem under consideration is normally phrased as a narrative (short story) that describes the sequence and effects of successive critical object interactions. In the loaded-spring context, this could be

> Several *springs* with given <u>load masses</u> and <u>spring constants</u> are <u>extended</u> and <u>released</u> from a *launcher*. The *trajectory* is <u>recorded</u> and then <u>plotted</u> by a *detector*.

The italicized nouns in the description then map to physical or abstract objects, while the underlined nouns and verbs are, respectively, the attributes and actions (e.g. variables or functions) associated with these objects.

The implementation of time presents significant subtleties. To model physical time evolution, every object should in principle experience time independently so that e.g. the spring oscillates and the detector periodically records the trajectory by polling the internal computer clock. Unfortunately, code in which two or more processes or "threads" access the CPU requires complex, "multithreading" facilities to manage e.g. resource contention and time-sharing among decoupled processes. Therefore generally the unique **main( )** program sequences time evolution. Alternatively, a "time-server" object can be responsible for sequencing interactions or different objects can control sequencing during disparate periods of execution.

7.3 Requirements specification

As in procedural programming, once a problem has been defined, the necessary resources, such as server or Internet access, mathematical libraries, etc., should be identified and the time, expense and complexity of the project estimated. To quantify the input and output data and identify possible error conditions, the graphical user interfaces (GUIs) of the program should be specified, either on paper or by employing the GUI builders contained in nearly all commercial C++ integrated development environments. Problem boundaries, which are parameter regions that yield incorrect or unphysical behavior, should be isolated. The response to each distinct error case should be elucidated, which generally requires additional problem descriptions such as

> When more than four springs are activated by the spring launcher, an error message will be displayed and the spring launcher will cease to function.

7.4 UML diagrams

While object-oriented programs are often formulated by first assembling a list of the properties of each class in the program on an individual card or sheet of paper, software engineering packages can be employed to automate the development process, c.f. *A First Course in Computational Science and Object-Oriented Programming with C++*. The objects and their interactions are represented as blocks and links from which skeleton code with empty function bodies is automatically generated. The final program is obtained by inserting code into each function body. The steps in this process are as follows.

(1) The problem definition and requirements specification are summarized by *use cases* that depict the general manner in which the system components interact with the external users (actors) and with each other.
(2) *Sequence and/or collaboration diagrams* establish the order in which functions are called (messages are sent) by the objects that comprise the system.
(3) A *class diagram* incorporates the objects and functions identified in the sequence/collaboration diagram into distinct classes.
(4) The class diagram is converted to C++ skeleton code by selecting a menu item.

7.5 Classes and objects

Following the problem specification, individual objects and their mutual interactions can be generated. Formally, an *object* corresponds to an entity that possesses internal variables containing data that define its particular state together with a set of functions that describe its behavior in response to external stimuli (messages). That is, an object embodies a concrete (e.g. a spring) or abstract (e.g. a time server, matrix or complex number) entity that has a unique name (identity),

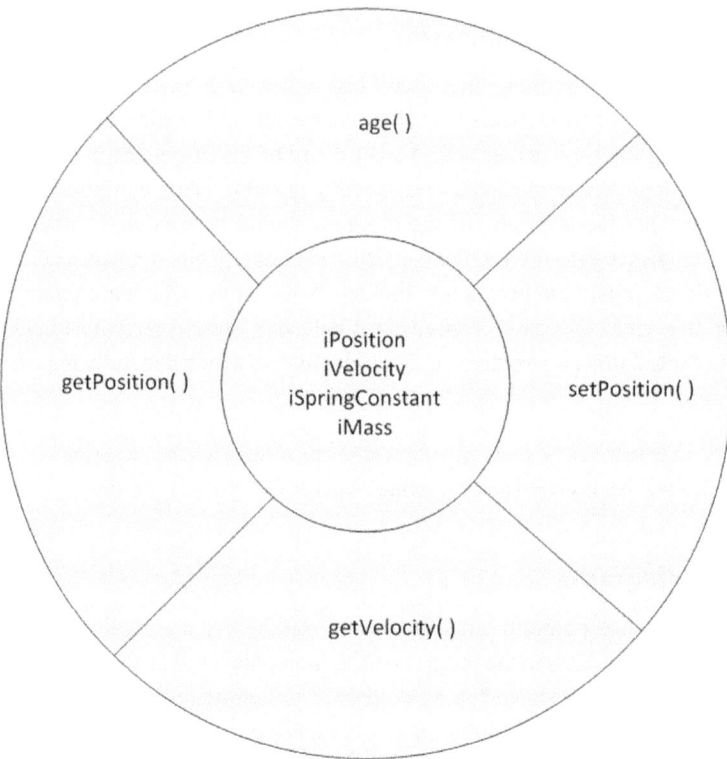

Figure 7.1

certain variables (properties) that characterize its state and, finally, a set of specific functions (behaviors). A *class* describes a group of all objects that share common properties and behaviors and provides a form (template) for the creation (instantiation) of these objects.

Normally the internal variables of a physical object are inaccessible except to the object. For example, the elements of a person's state such as his degree of thirst remain private, that is, accessible only to himself. Other people or objects can access these variables only by asking questions, i.e. directing messages to his public interface. Similarly, class components are private by default and can then be accessed or changed only by calling one of the public functions belonging to the class. Internal (private) variables can thus be thought of abstractly as located in an inaccessible core region, while public functions that govern the interaction of the outside world with the variables (the object's state) are situated in a user-accessible shell surrounding this region. Thus, a **LoadedMetalSpring** class for a spring with a certain spring constant that ages with time can be represented schematically as in Figure 7.1, in which internal variables with **get** and **set** functions can be externally viewed and altered, respectively.

An object is a specific realization of a class generated by assigning values to the variables present in the class diagram and specifying a unique object name. The class thus resembles a "form", "template" or "object factory"; filling in the entries in the form and associating the result with a name yields a unique class instance – the object. This parallels the manner in which e.g. an identification card (the object) can be procured by completing a form (the class) and specifying a unique name. An object can also be regarded as a generalized array variable that reserves memory and that can be assigned values by specifying its internal data members. From this perspective, a class corresponds to a generalized *type* (data structure) that collects variables with identical or different types together with related functions that transform these variables. This addresses the variable proliferation encountered in the procedural spring example since the variables of each spring are organized through their object name, as in, for example, **LoadedSpring LS1, LS2; LS1.iMass = 2, LS1.iPosition[0] = 3; ...**

7.6 Object discovery

As is evident from the above discussion, the number and type of objects in a program depend on the narrative employed to describe the physical problem, principally in the breadth of the narrative and in which nouns in the problem description are associated with classes as opposed to class attributes. For example, if the mechanism for launching the spring has a minimal number of attributes, then the spring in our problem description, rather than being assigned to a distinct **Spring** class, could be a component of a **SpringSystem** that combines the spring and the launcher.

The optimal strategy for designing classes associates each class with a single abstraction; that is, one definite and limited topic. Class members should further be orthogonal, in that the class interface provides the smallest number of independent public functions that enable the user to access all relevant behaviors of the underlying object. Further, these methods should be appropriate to the user as opposed to the programmer. As an example, a car that presents the user with two steering wheels, one for forward and a second for backward operation would lack desirability, although the inaccessible implementations (corresponding to **private** variables and functions of the **Car** class) for the two modes of operation differ substantially. Similarly, although the car design process might be simplified by a control that directly sets the motor RPM, this facility would not constitute a useful addition to the dashboard. Classes are sometimes additionally distinguished according to their functionality. Entity classes correspond to concrete physical entities or to abstract components, such as a **Vector**, that manipulate or store these entities. The interaction between external users and the system is handled by boundary classes typified by GUIs and plotting routines. Finally, control classes sequence the order of operations in the program.

Individual programming styles converge in code that maps every significant participating physical system to a separate object. However, grouping logically related functions into a few classes in order to decrease the overall programming effort instead yields a programmer-specific hybrid of object-oriented and procedural techniques. While these classes no longer represent the properties of individual physical objects, they still partition the independent variables and functions into structured, logical units.

To illustrate, in our spring example, a full description according to the problem description might be

```
Class LoadedSpring
Variables:
iLoadMass
iPosition
iVelocity

Class SpringLauncher
Variables:                              Functions:
iNumberOfSpringsToLaunch                launch( )

Class Trajectory
Variables:
iPosition[ ]
iVelocity[ ]

Class Detector
Variables:                              Functions:
iSpringTrajectory                       record( )
                                        plot( )

Class PropagationRoutines
Variables:                              Functions:
iSpring                                 propagate( )
iSpringTrajectory[ ]
```

However, in an abridged description, a **SpringSystem** class could provide a single generalized array of all data and methods that are logically related to the spring and its environment. This structures the variables and functions in the problem without requiring the coding of numerous classes. The **SpringSystem** class thus contains all functions and variables that influence or describe the spring, both before and after excitation, including **propagate( )** and **launch( )** functions and the coordinates of its trajectory. Eliminating the actual objects from consideration creates conceptual difficulties, since a spring does not launch itself or record its trajectory. However, the loss of clarity is often offset by decreased code size. If a **Graph** object reads and plots the trajectory information stored in a **SpringSystem** object, the latter class could be structured as

```
Class SpringSystem
Variables:                              Functions:
```

```
iLoadMass                 launch( )
iPosition[ ]              propagate( )
iVelocity[ ]
```

7.7 Inheritance

Inheritance enables specialized classes, termed derived classes, to be assembled from preexisting, more generic base classes. A derived class acquires the attributes and behaviors (internal variables and functions) of its base classes except for those that are added or explicitly redefined. Semantically, a derived class exhibits an "is-a" or a "kind-of" relationship to its base classes, as opposed to the "has-a" relationship that exists when a class contains an object of a second class as a member variable. Classes can inherit through several levels of derived classes. Elements that are shared by several derived classes should then reside at the highest applicable level of the inheritance tree.

To illustrate, consider the relationship between **LoadedSpring** and **LoadedMetalSpring** classes. If the latter class merely adds an **age( )** function to the **iPosition**, **iMass** and **iVelocity** variables of the former class, it can acquire these variables through inheritance. In public inheritance, public variables of the base class remain public in the derived class and can be directly accessed anywhere in the program. *Private* inheritance, in contrast, converts all inherited elements of the base class into private elements that cannot be accessed from outside the derived class. *Protected* inheritance converts base class elements into variables that are visible to and therefore can be inherited from the derived class by further derived class levels but are not accessible elsewhere in the program.

Chapter 8
C++ object-oriented programming syntax

Object-oriented design is followed by object-oriented analysis, the conversion of the design into code. This chapter summarizes central language features, since many aspects of object-oriented syntax rely on variable types that have not yet been introduced.

8.1 Class declaration

A class declaration informs the compiler that a certain identifier is associated with an user-defined class:

```
class Trajectory;
```

Once a class has been declared, the class name becomes functionally equivalent to any built-in type identifier such as **int** or **float**. Therefore, a compiler can subsequently resolve the function declaration:

```
void myFunction(Trajectory aT);
```

but has not yet been informed of the nature of the class components. Thus the second of the two statements below generates a compiler error:

```
Trajectory T;
T.plot( );                          // Error: T.plot( ) not defined
```

8.2 Class definition and member functions

A class definition, which must be terminated with a semicolon, specifies the elements of a class and can either include or omit the code (bodies) of its member functions. This appears to contradict the general rule that a variable is defined by the compiler as soon as sufficient information is available to determine its storage size in memory. However, the class definition merely leads to memory allocation for its internal variables together with the values of *the beginning memory addresses* of the locations at which the binary representations of the function bodies (the function records) are located. An example of a class definition in which the function body is not specified is (the semicolons in **void plot( );** and at the end of the class definition are both required)

```
class TrajectoryPlotter {
     public:
     float iPosition[100], iVelocity[100];
     int iNumberOfPoints;
     void plot( );          // This statement cannot be repeated!
};
```

The statement **void plot( );** cannot be repeated within the class definition, since at each occurrence the compiler reserves memory for an address of a function record. Repeating this statement would therefore incorrectly store this address in two locations.

The body (definition) of the **plot( )** function must subsequently be supplied once at global scope after the class declaration. The function prototype must be preceded by the class name **TrajectoryPlotter** followed by the *scope-resolution operator* **::** which indicates that this **plot( )** function belongs to the **Trajectory-Plotter** class, i.e.

```
void TrajectoryPlotter::plot( ) {
     metafl("XWIN");
     qplot(iPosition, iVelocity, iNumberOfPoints);
}
```

Supplying the body of the function within the class definition instead leads to (a semicolon is now not required after the **plot( )** function body)

```
class TrajectoryPlotter {
public:
int iNumberOfPoints;
float iPosition[100], iVelocity[100];
void plot( ) {
     metafl("XWIN");
     qplot(iPosition, iVelocity, iNumberOfPoints);
     }
};
```

A function whose body is supplied *within* a class definition is compiled as an inline function, with certain exceptions, such as if **for** loops are present.

Including function bodies in class definitions can lead to subtle problems. For example, in the program below, the class **First** must know that the class **Second** has a member variable **iSecond** before the body of its **print( )** function can be processed, while the class **Second** must similarly be aware that **First** possesses a member variable **iFirst** before its **print( )** function can be processed. Accordingly, for at least one of these classes, the body of the **print( )** function must be specified after the other class has been defined:

```
class Second;
class First;
// The compiler now knows Second and First are classes

class First {
    public:
    int iFirst;
    // so it can resolve Second here
    void print ( Second aSecond );
};
class Second {
    public:
    int iSecond;
    // and it can resolve First here
    void print ( First aFirst );
};
// The compiler is now aware of First's and Second's members

void First::print( Second aSecond ) {
    // so it can resolve aSecond.iSecond here
    cout << aSecond.iSecond << endl;
}
void Second::print( First aFirst ) {
    // and it can resolve aFirst.iFirst here
    cout << aFirst.iFirst << endl;
}
```

A very common error in writing code for classes occurs when one or more internal member variables are redefined within a member function as below:

```
class C {
    int iC;
    public:
    void setI (int aI ) {
        int iC = aI;                      // Error!
    }
};
```

The new variable **iC** hides the internal class member variable of the same name, so that calls to **setI( )** do not affect the internal member variable, which remains uninitialized.

8.3   Object creation and polymorphism

As stated earlier, a class, like a built-in variable type, can be viewed as a blank form (template) with entry fields that accept values. Each realization (instance) of the class is a separate copy of this form with a unique identifier (name) into which a set of (possibly random) values has been placed. A specific realization of the class constitutes a user-defined object variable just as an instance of a built-in type such as **int** is termed a built-in variable. *Defining an object with the type of a user-defined class is functionally identical to defining a variable of a built-in type such as int; thus a class implements a user-defined data type.* Accordingly, the syntax for the creation (instantiation) of objects coincides with that for built-in variables, for example,

```
TrajectoryPlotter TP1;
TrajectoryPlotter TP2;
```

generates two objects of type **TrajectoryPlotter**. *Since these statements are type declarations and do not call functions, parentheses do not appear* after **TP1** and **TP2** in the above statements. In the same manner as that in which **int m;** assigns a random value to **m**, the values of all internal variables in **TP1** and **TP2** above are random. For public variables, to replace the random bit pattern with a meaningful value, the *member-of operator*, **.**, which selects an element from the class with the name specified after the period, can be employed. Thus, values are assigned to the first two position variables, the first two velocity variables and the **iNumberOfPoints** variable in **TP1** through

```
main ( ) {
        TrajectoryPlotter TP1;
        TrajectoryPlotter TP2;
        TP1.iNumberOfPoints = 2;
        TP1.iPosition[0] = TP1.iVelocity[0] = 0;
        TP1.iPosition[1] = TP1.iVelocity[1] = 1;
        TP1.plot( );
}
```

While this code is placed inside **main( )**, it could equally well be located in any other function. The first two lines can also appear in the global space outside all function bodies after the **TrajectoryPlotter** class definition.

Classes implement polymorphism in a transparent manner because the behavior of a function is bound to the type of the calling object. For example, suppose the **plot( )** functions in the **Circle** and **Ellipse** classes plot a circle and an ellipse, respectively. Then, **Circle C1; C1.plot( );** draws a circle, while **Ellipse E1; E1.plot( );** instead draws an ellipse. This facility parallels real-world behavior, leading to concise and understandable code.

An object that contains only *public* members (and does not possess a user-defined constructor) can be conceptualized as a generalized array in which, unlike in a standard array, elements can belong to different types. These are accessed through the object name followed by the element name (possibly followed by an array index) instead of through an array index as in a standard array. Consistently with this paradigm, internal object variables can be assigned values at the point of definition following the same syntax as arrays. Thus the **iNumberOfPoints** variable and variables **iPosition[0]** and **iPosition[1]** of **TP2** are set to 2, 0 and 1, respectively, while all other variables are set to zero at the point of definition through the statement

```
TrajectoryPlotter TP2 = { 2, 0, 1 }
```

Regarding an object as a generalized array partially clarifies the result of assigning (equating) one object to a second object of the same type. Although such a manipulation is illegal for a built-in array, *each member variable of the second object, irrespective of access privilege (**public**, **protected** or **private**) is copied to the corresponding element of the first object,* as in

```
class C {
public:
      int iC[2];
};

main( ) {
      C C1, C2 = { 1, 2 };
      C1 = C2;
      cout << C1.iC[1] << endl;             // Output: 2
}
```

A similar element-by-element copying occurs when an object is passed to or returned from a function or when a class is initialized with a second class at the point of definition:

```
C C3 = C1;
```

which can also be written as

```
C C3(C1);
```

8.4 Information hiding

In C++ information hiding is governed by the **public**, **private** and **protected** keywords. When followed by a colon in class definitions, all subsequent class members acquire the specified access privilege until a new access keyword is encountered. *Internal variables and functions default to private* until the first occurrence of a **public:** or **protected:** keyword, since properties of a physical object are typically private. When a public or protected member of a class is accessed from outside the class, the member-of operator is generally required, as in

```
class InformationHidingExample {
      // private by default
      int iPrivateReadWrite;
      public:
      // set member function
      void setPrivateReadWrite( int aPrivateReadWrite ) {
            iPrivateReadWrite = aPrivateReadWrite;
            iPrivateReadOnly = aPrivateReadWrite *
                  aPrivateReadWrite;
      }
      // get member function
      int privateReadWrite( ) { return iPrivateReadWrite; }
      private:
```

```
      int iPrivateReadOnly;
      public:
      int privateReadOnly( ) { return iPrivateReadOnly; }
};

main( ) {
      InformationHidingExample IHE1;
      IHE1.setPrivateReadWrite( 4 );
      cout << IHE1.privateReadWrite( ) << endl      // output: 4
      cout << IHE1.privateReadOnly( ) << endl;      // output: 16
}
```

Consider first **iPrivateReadWrite**. No access keyword is specified before the declaration of this variable, which therefore defaults to private. However, the internal variable can still be assigned a value as in the third line in the **main( )** function through the public **setPrivateReadWrite( )** member function, which is termed a set member or writer function. Its value can be further accessed by the public get member or reader function **privateReadWrite( )**, as in the fourth line of **main( )**. A private internal class member with a get but not a set member function such as **iPrivateReadOnly** above is termed read-only, i.e. write-protected, from *outside* the class, while a write-only (read-protected) member contains a set but not a get member function. The naming conventions should be followed consistently. In particular, internal variables are composed of a prefix i followed by the actual variable name, set member functions prefix the variable name with set, while get member functions possess the name of the variable (very often, however, get member function identifiers are instead prefixed with gct). As well, observe that the variable **iPrivateReadOnly** appears in the **setPrivateReadWrite( )** function before its definition several lines later, seemingly violating the principle that, if the compiler has processed a program up to a certain token, it cannot access information appearing after this token. However, since functions appearing in a class definition are inlined, and calls to the class member function must follow the class definition, this apparent violation cannot occur in practice.

A common convention is to place the part of the class that the user can access, namely the **public:** interface, first in the class definition, followed by **protected:** and **private:** sections. The internal variables, which are normally protected or private, then appear at the end of the class body. (Often, constructors are additionally placed first in the definition, followed, if present, by the destructor.)

8.5 Constructors

Now consider the private read-only internal variable **iPrivateReadOnly** in the program of the previous section. Since this variable cannot be changed from outside the class, when an **InformationHidingExample** object is created, the variable contains a random value that can be accessed only by member functions of the class, such as **setPrivateReadWrite( )**. To set this variable instead to a meaningful value, such as zero, during the creation of the object, one might attempt to place an initialization statement

```
int iPrivateReadOnly = 0;                    // WRONG!
```

within the class body. While this procedure is valid in Java, C++ requires that internal member variables be initialized within *constructor* functions. These can possess any number of arguments, including zero, and are called at most once by each object at the point of definition. While e.g. print statements can be included in a constructor to facilitate debugging, in the final version of the program, a constructor should be used only for its intended purpose, namely to allocate and initialize class objects (or possibly for type conversion, as will be explained in Section 14.8).

A zero-argument constructor (which can equally well be a multiple-argument constructor for which default values are specified for all arguments) is labeled a *default constructor* and is invoked by a standard definition statement as in

```
InformationHidingExample I1;
```

Note the absence of parentheses after the object name. *C++ supplies a default constructor, which generally assigns random values to all internal class variables, only in the case that no user-defined constructors are present.* A class can contain any number of additional constructors, each of which must possess a unique sequence of arguments, differing in number and/or type. Since constructors are employed only to define objects, they do not possess a return type and are named identically to the enclosing class, which is consistent with the syntax of a definition statement, which similarly employs the class identifier.

Two non-equivalent options exist for initializing an internal variable in the body of a constructor function. In a zero-argument default constructor, both internal variables in the **InformationHidingExample** class can be initialized to zero, either within the constructor body,

```
class InformationHidingExample {
    ...
        InformationHidingExample( ) {
            iPrivateReadOnly = iPrivateReadWrite = 0;
        }                                    // Default constructor
    ...
};
```

or prior to the execution of the code in the constructor body through an *initialization list*:

```
class InformationHidingExample {
    ...
```

```
InformationHidingExample( ) : iPrivateReadOnly( 0 ),
      iPrivateReadWrite( 0 ) { }       // Default constructor
...
};
```

The initialization list, which is processed before the class object is actually constructed, must be employed to initialize e.g. **const** *internal variables, since, if the object were constructed prior to initialization, these variables would contain random bit patterns that could not subsequently be changed in the constructor body.*

A two-argument constructor could take the form

```
InformationHidingExample( aPrivateReadOnly, aPrivateReadWrite ):
      iPrivateReadOnly( aPrivateReadOnly ) {
      iPrivateReadWrite = aPrivateReadWrite;
}
```

However, when introducing any non-default constructors, one must be extremely attentive to the fact that *if the user supplies a constructor with any number of arguments, a default constructor is no longer automatically generated by the system.* That is, the compiler assumes that, if non-default constructors alone are present, the internal variables represent physical quantities that do not possess standardized values. However, *features in the code that at first sight do not seem to require a default constructor can fail if one is not present. Accordingly, a default constructor should always be defined if a non-default constructor is introduced.* For example, c.f. Section 8.8, if a class, **C**, contains an object of a second class, **D**, as an internal member variable, the default constructor of **D** will be called when an object of type **C** is constructed unless a non-default **D** constructor is present in the initialization list. For similar reasons, if a default constructor is absent from a base class, user-defined constructors must be supplied in all derived classes.

8.6 Examples

As a first object-oriented program, the following code defines a class **Rectangle** with two private double precision members **iLength** and **iWidth**, public set and get member functions for these two members, a two-argument constructor that sets the length and width to any two user-specified values and a void function **area( )** that computes and then prints the area. Subsequently, the main function creates a **Rectangle** with a length of 10 and width of 20 and displays its area:

```
#include <iostream.h>
class Rectangle {
      public:
            Rectangle( double aLength, double aWidth ) :
                  iLength( aLength ), iWidth( aWidth ) { }
            double length( ) { return iLength; }
```

```
                double width( ) { return iWidth; }
                void setLength( double aLength ) { iLength = aLength; }
                void setWidth( double aWidth ) { iWidth = aWidth; }
                void area( ) { cout << length( ) * width( ); }
        private:
                double iLength;
                double iWidth;
};
main( ) {
        Rectangle R1( 10, 20 );
        R1.area( );
}
```

The second program defines a **Vector** class that encapsulates (wrappers) **double** arrays to enable bounds checking. The private member variables of **Vector** are the double array **iArray[1000]** and the integer **iArraySize**, indicating the number of actual stored array elements. The public class functions called by **main( )** are as follows.

double getArrayElement(int aPosition) – returns the array element at **aPosition-1** or terminates the program through a call **exit(0)** of **stdlib.h** if **aPosition** is either less than 0 or greater than **iArraySize-1**.

int getArraySize( ) – get member function for the internal variable **iArraySize**.

void addLastElement(double aArrayElement) – calls **exit(0)** if **iArraySize** is 1000, otherwise adds the element **aArrayElement** to the end of the array and increments **iArraySize** by one.

void changeArrayElement(int aPosition, double aElementValue) – calls **exit(0)** if **aPosition** is larger than **iArraySize** or less than 0 and otherwise sets **iArray[aPosition]** equal to **aElementValue**.

Vector(double aArray[1000], int aArraySize) – a constructor that sets the first **aArraySize** elements of **iArray** to the corresponding values in **aArray** and calls **exit(0)** if **aArraySize** is less than 1 or larger than 1000.

void printArray( ) – a function that sends the values of all **iArraySize** elements of the array to **cout**.

```
#include <iostream.h>
#include <stdlib.h>
class Vector {
   public:
      Vector( double aArray[1000], int aArraySize ) {
         if ( aArraySize < 1 || aArraySize > 1000 ) exit( 0 );
            for ( int loop = 0; loop < aArraySize; loop++ )
               iArray[loop] = aArray[loop];
            iArraySize = aArraySize;
      }
      double getArrayElement( int aPosition ) {
            if ( aPosition < 0 || aPosition > iArraySize - 1 )
               exit( 0 );
            return iArray[aPosition];
```

```
      }
      int getArraySize( ) { return iArraySize; }
      void addLastElement( double aArrayElement ) {
          if ( iArraySize == 1000 ) exit( 0 );
          iArray[iArraySize++] = aArrayElement;
      }
      void changeArrayElement( int aPosition, double aElementValue ){
          if ( aPosition < 0 || aPosition > iArraySize - 1 ) exit( 0 );
          iArray[aPosition] = aElementValue;
      }
      void printArray( ) {
          for (int loop = 0; loop < iArraySize; loop++ ) {
              cout << iArray[loop] << endl;
          }
      }

   private:
      double iArray[1000];
      int iArraySize;
};

main( ) {
   double a[100] = { 1, 2, 4, 9 };
   Vector V1( a, 4 );
   V1.addLastElement( 10 );
   cout << V1.getArraySize( ) << " " << V1.getArrayElement( 4 ) << endl;
   V1.printArray( );
}
```

8.7 Wrappering legacy code

While many scientific programs exist only in FORTRAN or C versions, legacy
C source code can be rewritten as object-oriented code by wrappering related
routines into C++ classes, although slight differences in syntax between C
and C++ must sometimes be addressed. (Compiled FORTRAN code can also
be incorporated with some effort into C++ classes, see Appendix D of *A First
Course in Computational Science and Object-Oriented Programming with C++*.)
For example, to wrapper the procedural DISLIN graphics code

```
float x[1000], y[1000];
int numberOfPoints;
// .... Assign values to x, y, numberOfPoints
metafl( "XWIN" );
qplot( x, y, numberOfPoints );
```

into a class, functions related to plotting are packaged together with the data that
they process. That is, the arrays **x** and **y**, together with **numberOfPoints**, are
converted into internal class members. A constructor is supplied to transfer data

into the internal variables and a **draw( )** function introduced to plot these values
as follows:

```
class Graph { public:
    int iNPoints;
    float iX[1000], iY[1000];
    Graph( float aX[ ], float aY[ ], int aN ) : iNPoints( aN ) {
        for ( int loop = 0; loop < iNPoints; loop++ ) {
            iX[loop] = aX[loop];
            iY[loop] = aY[loop];
            }
        }
    void draw( ) {
        metafl( "XWIN" );
        qplot( iX, iY, iNPoints );
        }
};
```

Sets of data points can then be stored and graphed as follows:

```
#include <iostream>
#include <dislin.h>
using namespace std;
// insert Graph class here

main ( ) {
    float x[2] = { 0, 1 }, y[2] = { 1, 2 };
    Graph G1( x, y, 2 );
    y[1] = 3;
    Graph G2( x, y, 2 );
    G1.draw( );
    G2.draw( );
}
```

In contrast to the procedural case, two **Graph** objects rather than four disjoint
arrays are required in order to store two sets of position and velocity data.

8.8 Inheritance

As noted earlier, in object-oriented programming functionality in one class shared
by a second, derived, class can be automatically incorporated into the derived
class through inheritance. However, while a derived class inherits all the non-
private variables and functions of its base class that are not explicitly redefined
in the derived class, the access privileges of the class members can be altered
depending on the form of inheritance. If a class **D** inherits from a class **C** through
public inheritance, the access privileges of all public and protected elements in
the base class are preserved in the derived class unless an element is explicitly
redefined in the derived class with a different access privilege. A less commonly
employed form of inheritance is *private inheritance*, in which all public members
of class **C** become private members of class **D**. Public inheritance is implemented
by commencing the definition of class **D** with **class D : public C { ...** To recall

this syntax, observe that, since the elements of the base class **C** are constructed before those of the derived class **D**, **C** forms a type of initialization list for **D**.

A derived class constructor constructs its base class components by calling the default base class constructor *before* the derived class properties are constructed. A user-defined base class constructor can be employed in place of the default constructor but then must be placed in the initialization list of the derived class constructor. If the base class lacks a default constructor, a non-default base class constructor must be present in *every* derived class as illustrated below:

```
class D {
      public:
      int iD;
      D( int aI ) : iD( aI ) { }        // Note: default constructor absent
};
class C : public D {
      public:
      C( D aD ) { iD = aD.iD; }         // Error: initialization list required
      C( D aD ) : D( aD.iD ) { }        // OK
};

main( ) {
      D D1( 1 );
      C C1( D1 );
      cout << C1.iD << endl;            // Output: 1
}
```

Since a derived class is a specialized form of the base class, a derived class object is by definition also a base class object. That is, suppose a derived class "AirFilter" is a derived class of the base class "Filter". Clearly an AirFilter is a particular type of Filter, therefore it can be employed anywhere in the program where a Filter is expected. However, except as discussed in Chapter 14, the object then behaves as a Filter, rather than an AirFilter, as illustrated by the following generic example:

```
class C {
      public:
      void print ( ) { cout << "C ";}
};
class D : public C {
      public:
      void print ( ) { cout << "D ";}
};

main( ) {
      C C1;
      D D1;
      C CArray[2];
      CArray[0] = C1;
      CArray[1] = D1;          // Valid: D1 is also a C object!
      D1.print( );             // Output: D
      CArray[1].print( );      // Output: C
}
```

The compiler reserves an amount of memory appropriate to a **C** object for each element in the array **CArray**. Therefore, when a derived **D** object is placed into the array, the additional features that extend past this memory space are discarded, leaving only **C** properties. That the **C** properties of **D** are constructed first when a **D** object is constructed in fact implies that these properties occupy the beginning of the memory space assigned to the **D** object.

Since a private variable in the base class such as **iPosition** in class **C** below *is never accessible* in the derived class, the following code generates a compiler error:

```
class C {
    private:
    double iPosition;
    public:
    double iVelocity;
};
class D : public C { public:
    setPosition( double aPosition, double iVelocity ) {
        iVelocity = aVelocity;
        iPosition = aPosition;
        // Error -- iPosition is private to class C!
    }
};
```

8.9   The "protected" keyword

There are two standard procedures for accessing base class variables such as **iPosition** in the above program from within a derived class. The first is to supply public get and set member functions for the **iPosition** variable within class **C**; however, these will be accessible from anywhere in the program. Access to an internal base class variable can instead be restricted to derived classes through the *protected* keyword. A variable or function that is protected in a base class is accessible and remains protected in all public derived classes, but is inaccessible (private) to all other program units. Thus replacing **private:** by **protected:** in the second line of the above program resolves the compilation issue, since **iPosition** is then accessible from within both the base and the derived class, but not from elsewhere in the program. Substituting **protected** for **public** in the line **class D : public C {** instead implements *protected* inheritance in which both public and protected members of the base class become protected members of the derived class, while private members of the base class are by definition inaccessible from within the derived class.

8.10   Multifile programs

An executable program can be composed of several source and object files that are respectively compiled or linked together. Header files must, however, guard against repeated definitions and must insure that the program can be processed for any order of compilation. This generally requires appropriate preprocessor directives. If just one or a few files of a multifile program are typically changed at each program development step, compilation time can then be minimized by separately compiling each file into an object file. The object files are combined by passing appropriate linking options to the linker. Only files that are altered must subsequently be recompiled. Alternatively, compiled object files can be inserted into a program library. The elements of a program library are stored in an alphabetical lookup table that can be rapidly searched, decreasing link times.

Header files and program libraries can be understood through the following example in which a file **mySquare.cpp** contains a **main( )** program that calls two functions, **printSquare( )** and **printFourthPower( )**, each of which accesses a third function **square( )** in a **SquareCalculator** class. Accordingly, we introduce three header files with a .h extension that include function declarations and possibly class definitions, where, as noted above, header files that contain definitions must guard against multiple inclusion through **#ifndef ... #endif** and **#define** statements.

To develop this multifile program in Dev-C++, create a new console window project and type the first program labeled **square.h** below into the editor (overwrite the lines that are automatically generated when the project opens). Then select File → Save as, select from the drop-down menu in the Save as type entry field Header files, type for the File name **square** (do *not* add a **.h** to the file name here) and depress the Save button. This creates the file **square.h** in the project. Select File → New → Source File from the menu and depress the Yes button in the pop-up window labeled "Add new file to the current project?" and repeat the above steps for the header file **printsquare.h**. (Alternatively, the files below can be separately created in the program directory with a program editor and added to the project by repeatedly selecting the third to the last "Add to Project" icon on the upper toolbar.)

```
//square.h
#ifndef _square
#define _square
class SquareCalculator {
      int iValue;
public:
      SquareCalculator( int );
      int calculate( );
};
#endif

//printsquare.h
void printSquare( int );

//printFourthPower.h
void printFourthPower( int );
```

Next, the function definitions corresponding to each header file and the **main( )** program are each placed into three different code files with .cpp extensions by repeating the above steps except for saving the files as a C++ source file instead of a header file. Each .cpp file can include any number of the above header files, in particular

```cpp
//square.cpp
#include "square.h"
int SquareCalculator::calculate( ){ return iValue * iValue; }
SquareCalculator::SquareCalculator( int aValue ) :
       iValue( aValue ) { }
```

```cpp
//printsquare.cpp
#include <iostream>
#include "square.h"
using namespace std;
void printSquare ( int aM ) {
       SquareCalculator SC( aM );
       cout << SC.calculate( ) << endl;
}
```

```cpp
//printfourthpower.cpp
#include <iostream>
#include "square.h"
using namespace std;
void printFourthPower ( int aM ) {
       SquareCalculator SC( aM );
       cout << SC.calculate( ) * SC.calculate( ) << endl;
}
```

The main program is

```cpp
//mysquare.cpp
#include <cstdlib>
#include <iostream>
#include "printsquare.h"
#include "printfourthpower.h"
using namespace std;

main ( ) {
       int m = 4;
       printSquare( m );
       printFourthPower( m );
       system( "PAUSE" );
       return EXIT_SUCCESS;
}
```

The double apostrophes surrounding the name of an include file direct the compiler to search for the header file first in the user's current directory and afterwards in the compiler's include file subdirectories. Finally, depress the "Compile and Run" button to compile and run the project. This invisibly writes compilation and linker commands into a text *makefile* entitled **Makefile.win** that can be directly inspected in the program directory. The makefile determines which source-code files require recompilation so that only changed files and any new required files are recompiled by the "Compile and Run" command.

In multifile programs, **typedef**, **enum**, **const** and **inline** definitions are local to the file in which they reside and therefore should be placed in a header file that is included in each component file in which they are employed. A non-**const** variable defined in one file and accessed in a second file must be declared in the second file with matching type through the **extern** keyword (without an initializer, which would convert the declaration into a definition). For example, if we place a global variable **int M = 10;** in the file **mysquare.cpp** in the program above, it can be accessed in **square.cpp** only if the declaration **extern int M;** is present in this file. If the **extern** declaration statement appears inside a block in a given file, the variable **M** acquires the scope of the block.

8.11   const **member functions**

A function argument that is declared **const** cannot be changed within its body. Since variables passed by value to a function are isolated within the function from the remainder of the program, a function with either a **const** or a non-**const** argument can accept both **const** and non-**const** parameters from the calling program.

An internal member function of a class can additionally be prevented from modifying the class's internal data members by placing the keyword **const** between the end of the parameter list and its code body as in

```
class C {
    public:
    int iM;
    void print( const aN ) const { cout << iM * aN << endl; }
};
```

The compiler would then generate an error message if the **print( )** function alters either its argument **aN** or the internal variable **iM** of the **C** class. A function that cannot alter internal class members is termed a **const** member function. Get member functions should preferably be coded as **const** member functions.

If a **const** member function is overloaded by a non-**const** member function with the same name, the **const** member function is called if the object that invokes the function is declared **const** (indicating that its internal data members cannot be changed), otherwise the non-**const** member function is activated:

```
class C {
    public:
    int iI;
    void print( const int aJ ) const { cout << iI * aJ << endl; }
    void print( const int aJ ) { cout << ++iI * aJ << endl; }
};

main( ) {
    const C C1 = { 1 };
    C1.print( 1 );                          // Output: 1
```

```
        C C2 = { 1 };
        C2.print( 1 );                          // Output: 2
}
```

If the first **print( )** function is not present, a warning message is generated by the Dev-C++ compiler and the output **2 2** is obtained. While the **const** keyword is generally omitted below for brevity, its use can significantly reduce inadvertent programming errors.

Chapter 9
Arrays and matrices

In this and the following chapter, arrays, references and pointers are examined. These will prove essential to a more detailed discussion of objects and classes.

9.1 Data structures and arrays

An array constitutes a built-in (e.g. native to the compiler) data structure, which denotes a collection of related quantities with common ordering properties. Each object in an array must be of the same type and is accessible through an integer index. Other commonly occurring data structures include the following.

A *bag* – an unordered collection of objects.

A *set* – a bag in which no object can appear more than once.

A *list* – an object sequence that enables navigation between an object and its successor.

An *ordered list* – a list with the property that an object can be accessed through its position as well as by way of the neighboring object.

A *sorted list* – a list with elements stored according to a given ordering operation.

A *key set* – an object collection that employs a key such as a word or number to locate elements.

A *stack* – a container for which elements can be added or removed only at the first (top) position; that is, only the last element to be added is accessible at any given time.

A *queue* – a container constructed such that only the oldest (first) element in the container can be removed at any given time.

9.2 Array definition and initialization

Since the C++ compiler defines a variable once it can establish the amount of required memory, a (compiler-allocated) array is defined once the *array type and the number of array elements* have been specified, so that an appropriate amount of memory can be allocated. Since this memory allocation is fixed and *cannot be changed at runtime*, the array size must be an expression formed from positive integers and **const int** variables, such as

```
const int m = 3;
const int n = 4;
double a[2 * m * n]; // 24-element double array
```

The uninitialized array elements above contain random values. Were the array size not a **const int** it could be altered during program execution, as in

```
int m = 3;
cin >> m;
double a[m];          // compile-time error
```

A dynamic memory allocation procedure that enables such a procedure will be introduced later. However, unless the array size varies substantially for successive runs of the program, compiler allocation is more efficient and less error-prone.

As noted in Chapter 6, array elements can be initialized with the syntax, *which can be employed only when the array is defined*,

```
int a[3] = { 1, 2 };
```

The number of elements in brackets above must not exceed 3, the determining array size specified in the definition statement. If it is less than 3, the final elements are set to zero. Hence all elements in an array can be assigned zero values by

```
int a[3] = { 0 };
```

If an array is defined to be constant, its elements cannot subsequently be changed and must therefore be initialized at the point of definition through a statement such as

```
const int a[3] = { 1, 2, 3 };
```

Otherwise the array elements will contain random values that cannot subsequently (simply) be set to meaningful data.

9.3 Array manipulation and memory access

An array name in C++ possesses a fundamentally different interpretation than a variable name, as can be seen from

```
main ( ) {
     int a[2] = { 1, 2 };
     cout << a << '\t' << a[0] << '\t' << a[1] << endl;
}
```

which yields an output such as

```
0012FF84 1 2
```

The memory location displayed as the first value in the output indicates that the name of the **int** array variable, **a**, constitutes an alias (that is, an alternative name) for the address of the first array variable in the same manner as that in which a function name evaluates to the starting location of its instruction record.

To access a value stored in the array, the index operator **[ ]** is employed so that **a[0]** yields the **int** value stored at the starting location in memory, 0012FF84, in the above example. Similarly, **a[m]**, where **m** is an integer, addresses the **int** value stored at a position shifted from the starting location by **m** times the size of the memory occupied by a single element of the **int** array type. This implementation of the index operator explains the zero-based index of the n-element array, **a[0], a[1], ..., a[n − 1]** as well as the requirement that all array elements possess the same type. However, since the definition does not make reference to the array size, the index operator can access memory locations located at *any* positive or negative integer offset from a starting address. Consequently, an array can be initialized without specifying its size, as in

```
int a[ ] = { 1, 2, 3 };
```

which initializes **a** to a variable of **int** array type and allocates memory for three elements. That is, to be semantically precise, since **a[ ]** is interchangeable with **a[3]** in the above statement, the type of **a** in both cases is that of an **int** array, to which memory for three elements has been allocated, rather than a three-element **int** array.

Except in a definition statement where the purpose of the array index is not to address an array element but to determine the extent of memory allocation, the index of an array can be any **const** or non-**const** integer expression such as

```
a[m * ( n + 1 ) * 3] = b[n + 2];
```

However, such expressions must not evaluate to values beyond the array bounds (i.e. the limits of the memory allocated to the array).

A frequently occurring error is to assign an array to a second array,

```
a = b;                 // ERROR: a is a fixed address, not an lvalue.
```

rather than assigning the elements of **a** to the corresponding elements of **b**. Since **a** equates to the starting address of the array's memory, which is fixed during program execution and is therefore an rvalue, the above statement is rejected by the compiler. Instead the elements of the two arrays must be equated as in

```
for ( int loop = 0; loop < 10; loop++ ) a[loop] = b[loop];
```

When an array index equals or exceeds the array size established by the array definition, any one of three possible consequences can result. If the addressed memory location is located outside the memory space reserved by the operating system for the running program, the operating system intercepts the attempted illegal memory access, terminates the program and issues a "segmentation fault" message in a pop-up window, indicating that the program has violated its allotted memory segment. Since the array index then often greatly exceeds its permissible limits, the problem can usually be isolated either by printing out intermediate variables and viewing these in a debugger or by directly inspecting the source code.

In the event that the memory location accessed by the array variable instead occupies a region reserved for program operation but that is either not initialized or is associated with a variable of an incompatible type (e.g. a **double** instead of an **int**), the array element acquires a large, random value. The program then either generates unphysical results or raises an exception such as overflow or underflow, which, however, typically appears at a different code line than the actual error.

The third, and by far the most troublesome possibility, appears if the array index only slightly exceeds the array size and a second initialized variable of a compatible type is resident at the corresponding memory position, as in

```
main( ) {
     double b[2] = { 0.02, 6.0 };
     double a[2] = { 0.01, 3.0 };
     cout << b[0] << endl;
     cout << b[-1] << endl;
     cout << a[2] << endl;
     double wrongNorm = sqrt( a[0] * a[0] + a[1] * a[1] + a[2] * a[2] );
     cout << wrongNorm << endl;
}
```

which yields

```
0.02
3
0.02
3.00008
```

This error is indicative of many scientific calculations in which localized distributions such as pulses or particle wavefunctions that are small near their right and left endpoints are stored in adjacent arrays. The resulting unpredictable but minute changes in the program output can be misinterpreted as arising from e.g. discretization error in the numerical algorithm. Many compilers (but not Dev-C++) accordingly provide a switch that enables array bounds checking. Specialized programs can also be acquired that detect this and related memory errors.

9.4   Arrays as function parameters

Since an array is an rvalue, it cannot be returned by a function and assigned to a second array variable in the calling program. However, somewhat incongruously, a function can possess an array argument (since within the function an "array argument" is implemented as a constant pointer that *can* be equated to an array variable). If one of the elements of the array argument is altered within the function, however, the corresponding array element in the calling program similarly changes. The reason for this "pass (call) by reference" behavior is demonstrated by (note that the array size need not be included in the function argument, which requires only specification of the variable type)

```
void zero( int aA[ ] ) {
      cout << aA << endl;
      aA[0] = 0;
}

main( ) {
      int p[ ] = { 1, 2, 3 };
      cout << p << endl;
      zero( p );
      cout << p[0] << endl;
}
```

which yields

```
0012FF80
0012FF80
0
```

This indicates that the starting memory address of the array **p** is passed as a parameter to the function by the call **zero(p)**. Accordingly the starting memory location associated with the array argument **aA** is equated in the function to that of **p** in the calling program. Thus changing the value of an element of **aA** within the function body generates a corresponding change in **p** in the calling program. In this manner, memory for a copy of the array is not allocated in the function, avoiding substantial overhead for large arrays. It should here be remarked that a frequent error occurs when an array element is mistakenly employed in place of an array name as a function parameter in the calling program, e.g. replacing **zero(p)** by **zero(p[0])** in the above program.

Call by reference semantics connects variables across two disjoint blocks, compromising the intent of block structure. That is, if an array argument is assigned an unintended value in the function block, the corresponding parameter value in the calling block, which can be separated from the function by a large code distance, will change as well, leading to a concealed error that is difficult to detect.

9.5 Returning arrays as objects and object arrays

While a function cannot return an array, an object returned by a function and assigned to a second object transfers the values of all its internal variables to the latter object. Therefore, an array can be returned from a function if first wrapped in an object:

```
class Matrix {
     public:
     int iA[10];
};

Matrix f( ) {
     Matrix M1;
     M1.iA[0] = 10;
     return M1;
}

main ( ){
     Matrix B = f( );
     cout << B.iA[0] << " " << f( ).iA[0] << endl;
}                                               // Output: 10 10
```

The effective equivalence of user-defined and built-in types also enables arrays of objects to be defined in exactly the same manner as arrays of built-in variables. For example,

```
class Position {
public:
     double iPosition;
};

main ( ) {
     Position P1[2];
     P1[0].iPosition = 1;
     P1[1].iPosition = 2;
     cout << P1[1].iPosition << endl;          // Output: 2
}
```

9.6 const **arrays**

As stated earlier, a function argument that is declared **const** cannot be altered within the function body. Since standard variables are passed by value, the variable behavior is changed only within the function body; a change to the function argument does not affect the value of the function parameter in the calling program. For arrays and other argument types that are passed by reference, however, the **const** keyword also prevents unwanted side effects from occurring in the calling routine through an inadvertent change to the argument from within the function. As an example, in

```
void print( const int aM[ ], const int aN ) {
     cout << aM[0] << aN << endl;
}
```

attempting to assign a new value to any element of **aM** or to the variable **aN** within the function body yields a compiler error.

A non-**const** array can be passed as a parameter to a function with a **const** array argument. *However, a* **const** *array can be employed in a function call only if the corresponding array argument in the function is declared* **const**, so that

```
void print( int aB[ ] ) {
      cout << aB[0] << endl;
}

main ( ) {
      const int b[2] = {1, 2};
      print( b );                 // Error: aI not a const array
}
```

yields a compiler error since the function argument **aI** is not declared **const** in **print( )**. This results from the pass by reference semantics since otherwise the **const** property of the array in the calling program would be circumvented through changes within the function.

9.7 Multidimensional arrays

Arrays can be defined with arbitrary numbers of dimensions. As an example, a two-dimensional array with two rows and three columns that stores **int** values is defined as

```
int M[2][3];
```

The syntax, however, unfortunately reverses the significance of the array indices and should be construed as

```
(int[3])[2] M
```

indicating that **M** is an array of three-element integer arrays to which memory for two three-element arrays is allocated

The two-dimensional array **M** can be initialized at the point of definition in two ways. Since **M** represents an array of three-element integer arrays, its *first* argument (the 2) can be omitted in the initialization statement (unless memory is to be allocated for additional matrix rows beyond the ones specified in the initializer). Since each array in **M** is represented by elements enclosed in braces, where omitted elements are automatically set to zero, the matrix

$$M = \begin{pmatrix} 1 & 2 & 0 \\ 4 & 5 & 6 \end{pmatrix}$$

can be simultaneously defined and initialized by

```
int M[ ][3] = { { 1, 2 }, { 4, 5, 6 } };
```

where the elements in the first set of braces initialize the three-component array **M[0]**, while the elements in the second set initialize the components of **M[1]**. Alternatively, the initialization statement can be written as a single set of elements placed in the initialization list memory in the order in which they will appear in memory.

```
int M[ ][3] = { 1, 2, 0, 4, 5, 6 };
```

Finally, a $2 \times 3 \times 4$ three-dimensional matrix is defined as

```
int N[2][3][4];
```

9.8 Multidimensional array storage and loop order

That the definition **int M[2][3];** generates an *array of three-element integer arrays* each of which occupies 12 bytes of memory space can also be verified by noting that **sizeof(M)** returns 24 (6×4 bytes) while **sizeof(M[0])** is 12 (3×4 bytes). The elements of the three-element **int** array **M[0]** are **(M[0])[0] = M[0][0]**, **(M[0])[1] = M[0][1]** and **(M[0])[2] = M[0][2]**. These are followed in memory by the three-element array **(M[1])[0], ..., (M[1])[2]**. Accordingly, for the general case of an n-dimensional array, *rightmost indices vary most rapidly in memory*. In some other programming languages, such as FORTRAN and Octave/MATLAB, the leftmost indices instead vary most rapidly.

Of great importance to scientific applications is that this "row-major storage order" implies that the code

```
for ( int outerLoop = 0; outerLoop < n; outerLoop++ )
for ( int loop = 0, loop < m; loop++ )
    M[outerLoop][loop] = N[outerLoop][loop];
```

requires far less execution time than

```
for ( int outerLoop = 0; outerLoop < n; outerLoop++ )
for ( int loop = 0, loop < m; loop++ )
    M[loop][outerLoop] = N[loop][outerLoop];
```

In the latter case, memory is not addressed sequentially by the innermost, most rapidly varying, **for** loop (although modern compilers automatically optimize the loop order). For larger array dimensions, addressing successive elements can accordingly require accessing high-level cache memory rather than fast low-level cache or CPU memory registers. In the most unfavorable case, accessing neighboring elements induces a page fault in which portions of the matrix are exchanged (swapped) between the hard disk and fast memory, increasing execution times by orders of magnitude. Transversing array elements can therefore require careful planning. For example, two matrices **A** and **B** are normally multiplied as follows:

```
for ( int leftLoop = 0; leftLoop < leftDimension; leftLoop++ ) {
  for ( int rightLoop = 0; rightLoop < rightDimension; rightLoop++ ) {
    C[leftLoop][rightLoop] = 0.0;
    for ( int innerLoop = 0; innerLoop < innerDimension; innerLoop++ )
        C[leftLoop][rightLoop] += A[leftLoop][innerLoop] *
            B[innerLoop][rightLoop];
  }
}
```

Here the elements of **A** are traversed linearly (stride 1), but those of **B** are accessed suboptimally (stride **rightDimension**). If, however, **B** is replaced by its transpose, **BT**, the matrix product of **A** with **B** is obtained by replacing the innermost loop by

```
C[leftLoop][rightLoop] += A[leftLoop][innerLoop] *
                    BT[rightLoop][innerLoop];
```

insuring stride-1 access for both **A** and **BT**.

The following error is encountered surprisingly often when coding nested loops:

```
for (int outerLoop = 0; outerLoop < m; outerLoop++)
    for (int innerLoop = 0, innerLoop < n; outerLoop++)
        M[outerLoop][innerLoop] = N[outerLoop][innerLoop];
```

In the incrementation step of the second **for** loop **innerLoop** in has mistakenly been replaced by **outerLoop**, yielding an infinite loop.

9.9 Multidimensional arrays as function arguments

A multidimensional array can be employed as a function argument in the same manner as a single-dimension array. The one-dimensional array components of the multidimensional array can additionally be passed to functions of array arguments, as in

```
const int N = 2;
void f1( double aV[ ] ) {
    aV[0] = 1;
}
void f2( double aM[ ][N] ) {
    aM[0][0] = 2;
}

main ( ) {
    double M[N][N] = { 0 };
    f1( M[1] );                     // M[1][0] is now 1
    f2( M );                        // M[0][0] is now 2
    cout << M[1][0] << '\t' << M[0][0] << endl; // Output: 1   2
}
```

As with array parameters, the call **f2(M)** passes the multidimensional array name, which evaluates to its starting memory location, to **f2( )**. Therefore, **aM** in **f2( )** occupies the same memory space as **M** in **main( )**, resulting in pass-by-reference semantics.

Chapter 10
Input and output streams

C++ implements a device-independent approach to reading and writing data such that nearly the same interface (the public class members) is employed to send data to or receive data from different devices such as disk files, memory buffers and the keyboard and terminal. This is achieved by collecting functions and variables common to all input and output operations into an **ios** base class. Operations involving the standard input device (keyboard) **cin** are appended in a derived **istream** class, while functionality that enables data to be read from files and memory buffers is incorporated in the **ifstream** and **istrstream** subclasses of **istream**, respectively. Output operations are implemented in the same manner by the corresponding output classes, **ostream**, **ofstream** and **ostrstream**. The **iostream** class multiply inherits the internal functions and variables of both the **istream** and **ostream** classes. Finally, the **fstream** and **strstream** classes further specialize the **iostream** class by appending the properties required to both read and write to files and memory buffers (alternatively e.g. **fstream** can multiply inherit from **ofstream** and **ifstream**). Following the structure of the inheritance diagram, we first discuss the **iostream** class. Subsequently we discuss the interfaces to files and memory buffers associated with **fstream** and **strstream**.

10.1 The iostream **class and stream manipulators**

In C++, variables read into or written from a program initially reside in abstract objects termed streams, examples of which are **cin** and **cout**. These streams may be thought of as smart memory buffers enhanced through the inclusion of public, user-accessible functions that can act on the data stored in the stream buffer before its contents are processed by the program or sent to a device. The components that interface the memory buffer with the actual system devices are implemented as private functions and variables of the stream classes and are invisible to the user.

Stream objects, such as **cout** and **cin**, possess internal member variables labeled flags. Member functions of the stream access the flags, which control the stream's behavior. Flags relevant to **cout** govern in this manner, for example, the notation, precision and number of spaces employed in printing a value. (In reality, the flag value is typically assigned to a specific bit or bits of one or more bytes, which therefore can store several flags.) Flags that are common to all input and output classes are defined through an **enum** (set as a bitmask with members that evaluate to values 0, 1, 2, 4, . . .) in the **ios** base class and are addressed through the scope-resolution operator **ios::**. Important set member functions for flags and other internal member variables of the **cout** class are

```
cout.precision( n )           // n figures written out
                              // in floating pt.
cout.setf( ios::fixed )       // retains trailing zeros, turns
                              // off scientific notation
cout.setf( ios::left )        // left-justifies output
cout.fill( '.' )              // replaces all blank fields with a
                              // period
cout.setf( ios::scientific )  // scientific (floating point)
                              // notation
cout.width( m )               // reserves m spaces for output
```

Generally, if a property of the stream object is changed, the flag values are altered and the stream persists in the new state. The output width, set to **m** above, is, however, set to revert to its default value every time a new value is written, since data of a different type are generally extracted from the stream at the subsequent operation. Therefore

```
main ( ) {
        float f = .0001;
        cout.setf( ios::fixed );
        cout.setf( ios::left );
        cout.fill( '.' );
        cout.width( 20 );
        cout << f << endl;
        cout.precision( 20 );
        cout.setf( ios::scientific );
        cout << f << endl;
}
```

yields

```
0.000100 . . . . . . . . . . . .
9.999999747378751636e-05
```

The flags defined in **iostream** and its base classes can also be altered through stream manipulators defined in the header file **<iomanip>**. These alter the stream flag values when piped directly into an output stream:

```
#include <iomanip.h>

cout << setiosflags( ios::scientific );  // for scientific notation
cout << setiosflags( ios::showpoint );   // for decimal notation
cout << setw( n );                       // set width of output
                                         // field
cout << setf( ios::left );               // left-justify output
cout << setprecision( n );               // sets precision to n
                                         // digits after the
                                         // decimal point
```

Output can be sent directly to the printer by replacing all instances of the standard output stream **cout** by the printer stream **cprn**. Other standard streams are **cerr** (the "standard error" stream) and **clog**, both of which are associated by C++ with the terminal, and **caux**, which is bound to the serial communication port.

10.2   File streams

A file is an (often non-contiguous) allocated region on a storage medium such as a hard disk, floppy disk, memory card or CD-ROM. A file resides within a larger group of files termed a folder or (sub)directory and is accessed through either its absolute or its relative pathname. An example of an absolute pathname is **C:\Dev-Cpp\include\io.h**, which refers to the Dev-C++ include file **io.h**. The relative pathname to this file from the directory **C:\Dev-Cpp\bin** is **..\include\io.h**, where **..** is interpreted by the operating system as the parent directory, **C:\Dev-Cpp**, immediately above the current directory (a single period corresponds to the current directory). The relative pathname from the parent directory **C:\Dev-Cpp** is **include\io.h** (or, equivalently, **.\include\io.h**).

As stated earlier, by substituting **#include <fstream>** (not **ifstream** or **ofstream)** for **#include <iostream>** in a program, streams are generally defined that inherit the properties and behaviors of the **iostream** class but append the additional functionality required by file operations. In Dev-C++, however, **#include <iostream>** must be additionally present to access the standard output streams **cin** and **cout**, and **ofstream** and **ifstream** are often required in place of **fstream** for output and input operations. While file operations employ the same interface as **cin** and **cout**, they must be attached to a specific storage file location as in the definition statement

```
fstream myFileStream( "fileName" );
```

If the stream is employed exclusively for input or output, then **ifstream** or **ofstream**, respectively, can be substituted for **fstream**. Subsequently, data values are read from or output to the input or output file streams in the same manner as for **cin** and **cout**, e.g.

```
int r;
myInputFileStream >> r;
myOutputFileStream << setw( 8 ) << r << endl;
```

Assuming that **file.dat** containing only **int** values has been attached to the file stream **myFileStream**, data can be read up to the end of the file with either

```
while ( myFileStream >> r ) { ... }
```

since files, like strings, are terminated by a null character, or

```
while( !myFileStream.eof( ) ) { myFileStream >> r; ... }
```

where the **eof( )** member function of **fstream** returns true when the null termination character is reached and false otherwise.

When a file is opened for writing, its previous contents are normally deleted. To append data to a preexisting file, **ios::app** must be included in the stream definition:

```
ofstream fout( "out1.dat", ios::app );
```

The stream definition can also be separated from the process of attaching the stream to a file by introducing an open statement, as in

```
ofstream fout;
fout.open( "out1.dat", ios::out | ios::app );
```

After a subsequent close statement, the stream can be reassigned to another file,

```
fout.close( );
fout.open( "out2.dat" );
```

A file that is repeatedly opened in append mode, written to and then immediately closed can be copied during program execution, enabling output from a long-running program. All files are automatically closed by the operating system when a program terminates.

Storage space and access time can be reduced by storing information in binary as opposed to text (ASCII) format. This is accomplished by including the **ios::binary** specifier in either the **open** or the file definition statement as in **fstream fout("myFile.dat", ios::out | ios::binary);**. However, writing e.g. a **double** variable **a** then requires the somewhat cryptic syntax **fout.write((char *) &a, sizeof(a));**.

A stream such as **cin**, **cout** or a user-defined file stream such as **fout** above can be passed to a function. However, the function must act on the same file stream defined in the calling program, since a standard file stream cannot be copied. To implement pass-by-reference semantics and thus avoid copying, a function must employ an array, reference or pointer argument. Since only the first of these variable types has been introduced, the procedure is illustrated by passing an array of two file streams to a function:

```
#include <fstream>
void print ( ofstream aFstream[ ] ) {
     aFstream[0] << "Hello World" << endl;
}

main ( ) {
     ofstream fout[2];
     fout[0].open( "test1.dat" );
     fout[1].open( "test2.dat" );
     fout[0] << "Line 1 " << endl;
     print ( fout );
}
```

10.3 **The** string **class and** string **streams**

Many programs require the manipulation of textual information. A stream that
processes strings can be connected to a region of memory termed a buffer,
which is defined through the **strstream** class defined in **<strstream>** or the
corresponding **stringstream** class of the updated header file **<sstream>**. The
underlying container into which such string streams place data is a **string** object.
To create and manipulate individual **string** objects, the **string** header file must
be included. A blank string object is then created either by

```
string S;
```

or by

```
string S = " ";
```

A string can be initialized by

```
string S = "A string";
```

The operator + is overloaded for strings; for example, S + " : " + S yields the
string "A string : A string". Lexographic comparisons (comparison according to
dictionary order) are performed through the overloaded operators ==, !=, <, <=,
> and =.

Some important functions that act on string objects are

```
S.length( )
```

which returns the number of non-null characters in the string **S** such that the
length of a blank string is zero, and

```
S.erase( 3, 4 );          // result: A sg
```

which erases four characters in the string **S** starting at the position to the right
of the third character (writing simply **S.erase( );** or **S = "";** deletes the entire
contents of the string). A single element of a string can be accessed either through
S[m] or through **S.at(m)**, while the entire string can be converted into a standard
"C-type" character array terminated by the null character by employing **S.c_str( )**.
When **m >> length(S)**, **S[m]** returns the null character. Characters can be read
directly into or out of a string from a stream through the operators >>, and <<
and the **getline( )** function. The **find( )**, **replace( )** and **insert( )** functions can be
respectively employed to find a sequence of characters in a given string, replace
a character or a sequence of characters in a string with other values and insert
additional characters into a stream.

To illustrate the application of string streams, suppose that data, represented
below by the string "some input data", are to be written to a large number of
files labeled **output1.dat**, **output2.dat**, **output3.dat**, etc. Such names can be

generated automatically by piping the string "output" followed an integer and finally by ".dat" into a string stream as below:

```
#include <sstream> // Automatically includes
                   // string.h
#include <fstream>

main ( ) {
    char temp[10] = "input";
    string S;
    int numberOfFile = 1;
    stringstream inoutStream;
    ofstream fout;
    inoutStream << temp << numberOfFile << ".dat" << endl;
    inoutStream >> S; // Reads from string stream
    fout.open( S.c_str( ) ); // Converts string object to c string
    fout << "input data" << endl; // Data are placed in input1.dat
}
```

In the above program, a **string** object containing **input1.dat** is composed from two C-string variables and one integer variable by inserting the values into the string stream **inoutstream**. This object is then piped into the string **S** and converted through the **c_str( )** member function into a character array. Such a **char** array type can then be employed as the file name parameter of a call to **open( )**. While it might appear that the string stream could be replaced by **temp + numberOfFile + "dat"**, numeric variable types can be transformed in a simple fashion to strings only through insertion into a string buffer.

10.4   The toString( ) class member

Every class should possess a **toString( )** member function that returns a string containing properly formatted values and names of the internal class variables. If a class contains one or more user-defined objects as internal member variables, its **toString( )** function is assembled by calling the corresponding **toString( )** functions of the member objects. The example below demonstrates the application of string streams in this context. Here **ends** terminates an entire string while **endl** terminates a line within a string:

```
#include <strstream>
#include <iostream>
using namespace std;

class C {
    public:
    int iC;
    C( int aC = 0 ) : iC( aC ) { }
    string toString( ) {
            strstream s;
            s << "iC = " << iC <<ends;
```

```
            return s.str( );
    }
};

class D {
    public:
    C iC[2];
    int iD;
    D( int aD, C aC[ ] ) : iD( aD ) {
            iC[0] = aC[0];
            iC[1] = aC[1];
    }
    string toString( ) {
            strstream s;
            for ( int loop = 0; loop < 2; loop++ )
                    s << iC[loop].toString( ) << " ";
            s << endl;
            s << "iD = " << iD;
            s << ends;
            return s.str( );
    }
};

main( ) {
    C C1[2] = {1, 2};
    D D1( 2, C1 );
    cout << D1.toString( );
}
```

The output from **main( )** is

```
iC = 1 iC = 2
iD = 2
```

10.5 The printf **function**

Since the C++ language encompasses the C language with a few minor modifications, the standard C input and output routines, which conveniently format output
into columns, can be activated by including the **<stdio.h>** header file. Each line
is written to the terminal through the **printf( )** function whose argument contains line-formatting information followed by the output variables. Floating-point
numbers are formatted through the specifier **%m.nf**, where **m** and **n** are the numbers of integers to the left and right of the decimal point, respectively. An integer
is represented by **%id**, where **i** is the number of positions to be displayed, and a
string by **%ms**, where **m** is the number of characters in the string. Thus

```
#include <stdio.h>                // C I/O functions

main( ) {
    int m = 10;
    double x = 1.e2;
    char a[10] = "and ";
    printf( "x = %3.4f, %10s m = %10d", x, a, m );
}
```

sends the line

```
x = 100.0000, and m = 10
```

to the standard output device. A related function, **sprintf(buf, "x = %3.4f, %10s, m = %10d", x, a, m);**, instead places the output into a memory buffer such as **char buf[100];**.

Chapter 11
References

Reference and pointer variables, like arrays, store memory locations rather than values. The memory location of a reference cannot be changed and must be identified with that of a preexisting variable. Hence a reference can be handled in effectively the same manner as the variable it refers to. The address stored in a pointer can in contrast be changed at compile or run time.

11.1 Basic properties

A *reference* variable stores a fixed memory location of a preexisting compiler-defined variable (since any variable allocated by the compiler occupies a preset memory location during program execution). Because the address stored in the reference cannot subsequently be altered, the compiler can manipulate reference variables in the same manner as the variables they refer to. Accordingly, a reference variable on a functional level provides an *alias*, i.e. an alternative name or a nickname, for an existing variable. Numerous examples of aliases can be found in high-level computer applications; for example, a cell "A1" of a spreadsheet can be assigned an alias such as "grade" such that grade = 5 has the same effect as A1 = 5. In C++ reference-variable syntax

```
int A1 = 0;
int &grade = A1;
grade = 5;
cout << A1 << endl;                // Output: 5
```

Since for e.g. **grade** above to constitute an alternative name for **A1** all properties of the two must coincide, a reference *cannot be reassigned to a new variable* (since **A1** cannot be reassigned) and *must be initialized to a variable of the same type as in its definition*. That is, declaring **double &grade** = **A1** yields a compiler error. Additionally, since a reference is an alternative name for a *variable* it *cannot be initialized to a constant*. Memorizing the above will prevent countless programming errors.

11.2   References as function arguments

References are often employed to prevent the allocation of new memory space for function arguments. Recall that when it is invoked a C++ function allocates memory for its argument variables and subsequently copies the values of the parameters in the calling program into these newly reserved locations. Thus, changes to the values of the argument variables in the function body do not affect the corresponding function parameters in the calling program. However, in the case of array arguments the array name parameter that is copied into the new memory space for the function argument evaluates to the *address* of the memory allocated to the array rather than the *contents* of the array. The function therefore accesses and modifies the same memory locations through the index operator as the array parameter in the calling block.

Similarly, if the argument of a function is a reference variable, then, when the function is invoked, the argument is initialized to the fixed address of the calling parameter variable and becomes an alternative name for this variable. Hence, a change to the reference variable inside the function alters the parameter variable value in the calling block

```
void zero( int &aM ) { aM = 0; }

main( ) {
    int m = 4;
    zero( m );
    cout << m << endl;
}                       // Output: 0
```

Recall that when a function is invoked it first defines and initializes the variables in its argument list. Thus **zero( )** above implicitly initializes the reference according to

```
int &aM = m;
```

For this reason, *the parameter passed to a function reference argument cannot be a constant or a variable of a different type*, so that the following yield compile-time errors:

```
double d = 4.0;
zero ( d );             // Error: an int reference cannot
                        // be assigned to a double.
zero ( 4 );             // Error: a reference variable cannot
                        // be assigned to a constant.
```

This feature can be leveraged to insure that a function accepts only array parameters of a certain size and type. In particular, if a function signature contains a reference to an array with e.g. 20 **int** elements, as in

```
void print( int (&aI)[20] ) { cout << aI[6] << endl; }
```

only 20-element **int** arrays can be passed to **print( )** from the calling routine.

11.3 Reference member variables

If different physical objects share a common component, the object corresponding to the component should be included in each object as a reference member variable. As for **const** internal variables, c.f. Section 8.5, reference member variables must be contained in the initializer list of each class constructor since otherwise they would be constructed with the class and could not subsequently be assigned to a target variable. For example, suppose that several **PositionSensor** objects share a single **Printer** subcomponent so that a change to the Printer settings affects the behavior of all the objects identically. This can be modeled by creating a single **Printer** object in e.g. the **main( )** function and including this object by reference in each **PositionSensor**. Schematically, we have

```
class Printer {
      public:
      int iPrinterSetting;
};

class PositionSensor {
      public:
      Printer &aPrinter;
      PositionSensor ( Printer &aPrinter ) : iPrinter(aPrinter){ }
};

main( ) {
      Printer P1 = {3};
      PositionSensor PS1( P1 );
      PositionSensor PS2( P1 );
      P1.iPrinterSetting = 5;
      cout << PS1.iPrinter.iPrinterSetting << " " <<
              PS2.iPrinter.iPrinterSetting << endl; // Output: 5 5
}
```

A change to the properties of the **Printer** object **P1** by any enclosing object or by **main( )** simultaneously propagates to all objects that contain **P1** as a reference variable.

11.4 const **reference variables**

If a reference variable is declared **const**, the value of the variable that it refers to cannot be changed by assigning a new value to the *reference* variable. A **const** reference to a non-**const** variable resembles in the spreadsheet analogy a second name for a spreadsheet cell that permits only read access; i.e. the cell contents can be viewed but not modified by the owner of the name. However, a non-**const** reference generally cannot be initialized with a **const** variable, since this would contradict the intention of the original **const** variable definition, just as a non-**const** array parameter cannot be passed into a function with a **const** array argument. Hence the first reference definition below yields a compile-time error (in some compilers a warning message), whereas the second is correct:

```
const int m = 4;
int n;
int &k = m;                    // Compile-time error or warning
const int &l = n;              // Valid
```

A non-**const** reference can always be assigned to a **const** variable through a **const_cast** operation as in **int &p = const_cast<int &> (m);**. However, **p** is only assigned the value of **m**; subsequently changing **p** does not affect **m**, which remains **const**.

Assigning a **const** reference to a non-**const** variable avoids the overhead of copying function arguments, while still preventing changes to these arguments from within the function body. A **const** reference variable in the calling block cannot, however, normally be passed to a function with a non-**const** reference argument – again since the function implicitly defines and initializes the non-**const** argument with the **const** parameter. Thus, the second function call in **main( )** below generally yields a compile-time error:

```
void pr ( const int &aN ) { cout << aN << endl; }
void pr2 ( int &aN ) { cout << aN << endl; }

main ( ) {
    const int n = 3;
    pr( n );                   // Output: 3
    pr2( n );                  // Error
}
```

11.5   Reference return values

If a function returns a reference such as **int & myFunction( )**, the memory location associated with the return variable *inside* the function will be identical to the memory location of the variable to which it is assigned *outside* the function in the calling program. The function can then be employed as either an rvalue or an lvalue. In the latter case the function must be assigned to a *variable* of the same *type* as the return variable.

If a function reference argument is also returned as a reference variable, the function can be employed either as an rvalue to return the same, possibly modified, variable employed as its argument or as an lvalue to pass back to the program through its argument the same variable as that to which it is assigned. This facility is frequently employed in object-oriented programing (generally in combination with the **this** pointer) to simplify program syntax. To illustrate, the following nested chain of calls writes different messages to a single file. Since the function returns the same **ofstream** object as that which enters through its argument, two nested calls can be made to **print( )** without incurring an illegal copy operation (**fstream** does not provide the required operators to enable copying):

```
#include <fstream.h>

ofstream& print( ofstream& aStream, char aMessage[80] ) {
      aStream << aMessage << endl;
      return aStream;
}

main ( ) {
      ofstream myStream( "output.dat" );
      print( print ( myStream,
        "This is the first line" ), "This is the second line" );
}
```

Chapter 12
Pointers and dynamic memory allocation

A pointer variable like a reference or array variable stores a *memory address*; however, this address can be arbitrarily changed, enabling the contents of any accessible memory location to be addressed and manipulated directly. While enabling access to all available resources, new and subtle types of errors arise. For example, when a program requests additional memory during runtime from the operating system, the address of the starting location to the new, dynamically allocated variable is returned. Since the value stored in a preexisting compiler-allocated pointer variable can be altered, the running program can preserve the address passed back by the operating system. However, if the pointer variable was defined within an inner block, it will be destroyed when the block terminates. The location of the dynamically allocated memory is then lost and the memory cannot subsequently be accessed or later freed.

12.1   Introduction to pointers

A definition in C++ establishes the amount of memory space required for a variable and the interpretation of the value stored at this memory location. The value of a pointer variable is associated with the *starting memory address of a variable of a specified type, i.e. the value of a pointer is a memory address.* That is, a **double** pointer that stores a value such as 80000 interprets the 8-byte region from physical memory location 80000 to location 80007 as the storage location of a **double** variable. *The amount of memory space reserved by any pointer variable equals the number of bits required to store a hardware memory address* – on a 32-bit machine, an address requires 4 bytes so that *applying the* **sizeof( )** *function to any pointer yields 4.*

To define a pointer to a variable of e.g. type **int**, the notation **int *** is employed, where a pointer type is designated by a star. To define multiple pointer variables within the same statement, a star must precede each pointer. Thus the following defines **j** to be an **int**, while **pI** and **pK** are pointers to integers:

```
int *pI, j, *pK;
```

12.2 Initializing pointer variables

Although **pI** and **pK** are defined by the statement above, they are not initialized. Therefore, the 4-byte memory space allocated for these variables (on a 32-bit machine) contains a random bit pattern and the variables are therefore said to point to random memory locations. These pointers must be reassigned to the memory addresses of allocated **int** variables before they can be safely employed. For example, recalling that C++ by default manipulates memory addresses as hexadecimal constants, which are denoted by a preceding 0x or 0X, if the starting memory address of an integer variable is 0x0012FF88, then, after casting this constant value to the type of an **int** pointer, one can write

```
int *pM = (int *) 0x0012FF88;          // 0x = hexadecimal
```

A hexadecimal value can also be read in from the keyboard (in this case not preceded by 0x) through the following syntax (if the **hex** manipulator is omitted, the integer equivalent of the hexadecimal value should instead be entered):

```
int m, *pM;
cin >> hex >> m;                       // Input: 0012FF88
pM = (int *) m;
```

This procedure is of limited value since, except for those associated with physical devices such as a sound card or serial port, most memory locations in a program are specified as offsets from a starting location determined by the operating system at runtime.

12.3 The address-of and dereferencing operators

Address-of operator. The memory location of a variable is accessible during program execution through the *address-of* operator **&** (the ampersand does not possess the same meaning as the ampersand employed to define reference variables). To illustrate,

```
int m;
cout << &m << endl;                     // Output: e.g. 0012FJ4A
```

displays the address of the variable **m** on the terminal. This address can be stored in a preallocated **int** pointer variable:

```
int *pM = &m;                          // pK now contains 0012FJ4A
```

Unlike a reference, this memory address can be later reassigned to the address of a different variable as in the lines below:

```
int 1;
pM = &1;
```

Dereferencing and member-selection operators. Once a meaningful memory address has been stored in a pointer, the *value* at this address can be altered through the pointer variable. The value is accessed through the *dereferencing operator*, *, as follows:

```
int m = 4;
int *pM = &m;
*pM = 5;
cout << m << ' ' << *pJ;          // Output: 5 5
```

Thus ***pM** constitutes an *alias* for the *variable* **m**, the **int** variable located at the value of **pM**. Further, since ***(&m)** yields the same variable as **m**, the address-of and dereferencing operators comprise inverse operators. The variable **pM** can also be dereferenced through **pM[0]**, which possesses the same meaning as ***pM**.

A special form of the dereferencing operator exists for objects. Consider

```
class C {
     public:
     int b;
     void f( ) { cout << "test" << endl; }
};
```

A pointer **pC** to an object, **C1**, of type **C** is defined in **main( )**:

```
main( ) {
     C C1 = { 1 };
     C *pC = &C1;
     cout << (*pC).b << endl;   // Output 1: parenthesis required!
     pC -> f( );                // Output: test
}
```

In the expression **(*pC).b** appearing within the third statement of **main( )**, **pC** is deferenced to yield an alias for the object variable **C1** to which **pC** points and the member-of operator, **.**, is then applied to access the public member variable **b** of **C1**. The *parentheses are required* around ***pC** since the precedence of the member-of operator is higher than that of the dereferencing operator. Since objects are often manipulated through pointers in this manner in order to avoid the overhead associated with copying large blocks of memory, **(*pC).f( )** is abbreviated by a second member-selection operator, **->**, in **pC->f( );** on the fourth line of **main( )**. This circumvents the frequent error of omitting the parentheses around the dereferenced object in the former expression.

12.4   Uninitialized pointer errors

As noted in the introduction, since pointers can address any memory location permitted by the operating system, new classes of errors arise. A particularly common runtime error results when a value is assigned to an uninitialized pointer, as in

```
int *pI;              // pI can point anywhere in memory
*pI = 20;             // Typically forbidden by O/S
```

Here the pointer variable **pI** contains the address of a random memory location that statistically is almost certain to be located outside the memory space allocated to the program by the operating system. Placing a value into this location therefore raises a general protection fault leading to an operating-system runtime error message. Accordingly, *every uninitialized pointer or pointer that is detached from a meaningful memory location should be assigned the address* **0** or equivalently **NULL**. A logical test then insures that each pointer is initialized before dereferencing:

```
int *p = NULL;
cout << p << endl;       // Output: 00000000
... additional lines such as int l; p = &l; ...
if ( p ) *p = 3;         // Insures p is initialized
```

12.5 The const **keyword and pointers**

Since a pointer permits changes both to its value, which is the memory address that it stores, and, through the dereferencing operator, to the value stored at this address, three possible implementations of the **const** keyword exist for pointers.

Constant pointers. If **const** is placed after the * symbol in a declaration as in

```
int m = 1;
int * const pM = &m;
```

then the address that **m** stores (points to) is fixed, but the **int** value at this address can be changed through the dereferencing operator. Indeed, reading the definition from right to left indicates that the value of **pM** (the address stored in **pM**) is declared **const** through **(const pM)**, while dereferencing this address through *(const pM) yields a non-constant **int** variable. Since the address of the pointer cannot be changed, **pM** *must* be initialized to a sensible address when defined. For the above definition, we have

```
int k = 2;
*pK = 30;                // VALID, the int value can be changed.
pK = &k;                 // ERROR, the address in k is constant.
```

Note the similarity of this construct to a reference variable. Since a constant pointer cannot be detached from the variable to which it points, the only meaningful operations that can be performed on the pointer require dereferencing to access the underlying non-constant variable value. Thus, a reference constitutes a constant pointer that the compiler invisibly dereferences, transforming its syntax into that of the underlying variable.

*Pointer to a **const***. A second use of **const** as applied to pointers is given by

```
const int *pK;
```

or equivalently

```
int const *pK;
```

As the declaration indicates, the value ***pK** located at the address held in **pK** is a **const int** in the sense that it cannot be changed *through* **pM**, but the value of **pK** (the memory location stored in the pointer) can be altered. That is,

```
int k = 1;
pK = &k;              // OK: address held in k can be changed.
*pK = 4;              // Error: k == 1 cannot be changed through pK
```

*Constant pointer to a **const***. Analogously to a **const** reference variable, for the declaration

```
const int * const pM = &m;
```

the memory location stored in the pointer variable cannot be altered, while the value stored at this location cannot be modified through the pointer variable.

12.6 Pointer arithmetic

Suppose that the memory allocated to a variable of type **variableType**, **sizeof(variableType)**, is N bytes, where **variableType** can be any built-in or user-defined type. Then, if e.g. **variableType *pK = (variableType *) 00000040;**, ***(pK + m)** corresponds to the value of the variable of type **variableType** stored at starting address $40 + m * N$, which is identical to **pK[m]**. Other arithmetic operations on pointers are implemented similarly.

12.7 Pointers and arrays

Since an array variable evaluates to a constant memory address, it resembles a constant pointer and can even can be assigned to a pointer of a non-array type. However, an array definition differs fundamentally from a pointer definition since **int a[10];** *allocates memory for* 10 **int** *variables*, while **int * const pA;** *only allocates memory for a single pointer variable that stores the memory address of an* **int** *variable* as illustrated below:

```
int a[3] = { 1, 2, 3 };
cout << a << endl;                              // Possibly: 001H2B00
cout << a + 2 << '\t' << *( a + 2 ) << endl; // Output:001H2B08 3
int *p = a;
cout << p[2];                                   // Output: 3
```

12.8 Pointer comparisons

When comparing two values through pointers to these values, the comparison operator must be applied to the *dereferenced* pointers, not to the pointers themselves. That is, in

```
int *pA = &m, *pB = &n;
if ( pA == pB ) cout << "same addresses" << endl;
if ( *pA == *pB ) cout << "same values" << endl;
```

the first **if** statement compares the *values* of the pointer variables **pA** and **pB**, which are the *addresses* of the integer variables that **pA** and **pB** point to, and thus determines whether **pA** and **pB** point to the same variable. While this could be intended, generally the objective is to compare the values of the *variables* which **pA** and **pB** point to. The pointers must then be dereferenced before effecting the comparison as in the second **if** statement above. A similar error occurs if the index operator, [ ], is omitted in comparing two array variables. Since each array occupies a unique, compiler-assigned, memory location, the logical condition will then always evaluate to false.

12.9 Pointers to pointers and matrices

Consider a pointer variable that stores the memory address of a second pointer, such as a pointer to an **int**. Just as the type of a pointer to an **int** is denoted **int ***, a pointer to an **int** pointer is defined by **int **ppN**. Again this definition is interpreted from right to left: applying the rightmost dereferencing operator * to **ppN** yields a variable of type **int ***; dereferencing this variable one more time is equivalent to dereferencing **ppN** twice and yields an **int** value, e.g.

```
int n = 4;
int *pN = &n;
int **ppN = &pN;
cout << **ppN << ' ' << *pN <<endl;          //Output: 4 4
```

Hence **pN[0]** is equivalent to ***pN** or **n**, while **ppN[0]** equates to ***ppN**, the pointer **pN**. By extension, **(ppN[0])[0]** or **ppN[0][0]** coincides with ****ppN**. A two-dimensional array can thus be constructed from a two-element array of pointers or as a pointer to a pointer:

```
int a[3] = { 1, 2, 4 };
int b[3] = { 3, 5, 6 };
int *pA[2];
pA[0] = a;
pA[1] = b;
int **ppP = pA;
cout << ppP[0][2] << '\t' << pA[0][2] << endl; // Output: 4 4
```

Note that, unlike for a matrix, the rows of **ppP** do not necessarily contain the same number of elements.

12.10   String manipulation

A C-string, often simply termed a string, is formed from a character array terminated by a null character that enables C++ functions and operators to cease processing under the control of a logical condition. C-strings can be initialized when declared through either

```
char s1[6] = "Hello";
```

or

```
char *pS2 = "World";
```

The second procedure reserves space for the terminating, null character in the string "World"; however, the size of the character array (6 bytes) allocated to **pS2** cannot be manipulated and would prove insufficient if a longer string were later stored in **pS2**.

The insertion and extraction operators can be employed to read a character string from the keyboard, or to write a string to the terminal. However, to preserve whitespaces in an input line, the **getline(s1, i1)** function can be employed to read the first **i1 - 1** characters of the line up to a carriage return, including whitespace characters. The function then appends a null character and stores the result in the array **s1**, which therefore must be at least **i1** characters long. Alternatively, **getline(s1, i1, 'c')**, reads **i1 - 1** characters from the input stream up to the termination character, **c**. That is,

```
char s[10];
cin.getline( s, 10, '|' );
cout << s;
```

yields the output "A Test" for an input line

```
A Test|
```

C++ provides native (built-in) string-handling functions that act either on character arrays or on pointers to character arrays. The following code computes the length of **s1** above, copies the first two letters of **pS2** into **s1** and compares the first two letters of **s1** and **pS2**:

```
cout << strlen( s1 );            // Output: 5
cout << strncpy( s1, pS2, 2 );   // Output: Wollo
cout << strncmp( s1, pS2, 2 );   // Output: 0 (s1 is now Wollo)
```

The output of **strncmp(s1, s2, n)** is 0 if the first n characters coincide, -1 if **s1** precedes **s2** in lexical (dictionary) ordering and $+1$ otherwise. In **strncpy( )**, which copies the first n characters of **s2** into **s1**, **s2** must be the same size as or smaller than **s1** in order to prevent a runtime memory allocation error. The functions **strcpy( )** and **strcmp( )** function identically but do not accept a third argument.

Identical operations can be performed on the corresponding **string** class objects introduced in Section 10.3 as follows:

```
#include <string>

main( ) {
    string S1 = "Hello";
    string S2 = "World";
    cout << S1.length( ) << endl;                  // Output: 5
    cout << S1.replace( 0, 2, S2, 0, 2 ) << endl; // Output: Wollo
    cout << S1.compare( 0, 2, S2, 0, 2 ) << endl; // Output: 0
}
```

The member functions **begin( )** and **end( )** return pointers to the initial and final elements of a **string** object. This facilitates iterating over the characters in a string, as in

```
string S = "a string";
for ( char* p = S.begin( ); p != S.end( ); p++ ) cout << *p;
```

which writes **S** to the terminal.

12.11 Static and dynamic memory allocation

Static memory allocation. Memory for variables and arrays that are allocated when the compiler processes definitions is termed *statically allocated*. The compiler also issues instructions to deallocate memory at the point at which a variable will no longer be used (generally when the block containing the definition is terminated). Thus in

```
float a[10];
void fun( ) { int b[10]; }
void main( ) { int c[10]; { int d[10]; } }
```

the *lifetime*, or the period during which memory is reserved, of the *global* array **a** extends from the beginning to the end of the program while the lifetime of the *local* arrays **b**, **c** and **d** extends from their definition statements to the end of the encompassing block.

Dynamic memory allocation. Despite the efficiency of compiler memory management, the amount of memory required by a program often depends on the outcome of a user interaction or a logical condition evaluated during execution. By postponing memory allocation until runtime, a program can exploit the resources of large computers while still executing successfully on small machines. For example, a computer game that reserves memory for a new player upon request should function on any system, but the maximum number of players will necessarily be limited by the available hardware.

In C++ a running program can request memory from the operating system for a variable of a certain type by invoking the **new** operator. This operator returns the starting memory address of the *dynamically* allocated space as a temporary *pointer* with the type specified in the **new** statement. To preserve the address so that the space can be accessed later by the program, it must be assigned to a compatible pointer variable previously allocated by the compiler in a definition statement. That is,

```
int *pM;          // Compiler allocated pointer to an int
pM = new int;     // Memory is allocated for an int
                  // and its address is stored in pM.
```

These two steps are often combined into the single statement

```
int *pM = new int;
```

A dynamically allocated built-in variable can be initialized when defined

```
double *pD = new double(100.0);
```

The syntax for the dynamic allocation of an *array* of 10 doubles is instead

```
double *pD = new double[10];
```

Clearly, *parentheses for initialization should not be confused with square brackets for array allocation.* If a request exceeds the maximum amount of available memory, the new operator either throws an exception or, in older compilers, returns a null pointer (a pointer with a value of 0). (This behavior can be forced in modern compilers by including the <**new**> header file together with the statement **new_handler set_new_handler(0);**.) Since dynamic memory is allocated by the operating system at runtime, it can be controlled by the running program as below:

```
int *pM;
cin >> n;
if ( n > 0 ) {
     pM = new int [n];
     if ( pM == 0 ) cout << "out of memory";
}
```

The lifetime of memory allocated through a call to **new** *extends until the end of the program unless it is deallocated with a* **delete** *or* **delete [ ]** *statement. The second of these statements must be employed whenever an array is allocated as in*

```
int *pM = new int;          // Memory allocated for int variable
delete pM;                  // Memory deallocated
pM = 0;
if ( !pM ) pM = new int[10]; // Memory allocated for int array
delete [ ] pM;              // Memory deallocated
pM = 0;
delete pM;                   // No effect
```

The square brackets after **delete** in the fifth line insure that the memory associated with the entire array will be deleted. If the brackets are omitted, only
the memory at the position of the first array element is certain to be deleted,
possibly yielding a memory leak. In contrast, **delete [ ]** can be safely applied
to non-array variables. The **delete** or **delete[ ]** statements do not affect the program when applied to null pointers. Note that the compiler-allocated pointer **pM**
remains allocated until the block in which it is defined terminates – the **delete**
statement *frees the memory which* **pM** *points to but does not delete* **pM** *itself.*
Accordingly, following a **delete** statement the value of **pM** can still be set to the
memory address of a different variable or to a new dynamically allocated memory address. Whenever a dynamically allocated variable is deleted, all pointers
to this variable should be set to the null pointer so that assigning a value to
the dereferenced pointers yields a runtime error instead of accessing deallocated
memory.

12.12 Memory leaks

The **new** operator allocates memory through an operating system call and returns
an address to this memory. As long as the program retains this value, it can
manipulate the contents of, or deallocate, the dynamically allocated memory.
While the address could be stored in any memory device, it is normally assigned
to one or more compiler-assigned pointer variables. If all these pointers are
destroyed when the blocks within which they are defined terminate or their values
are overwritten, the address to the allocated memory is irrevocably lost. The
dynamically allocated memory, while still reserved by the operating system for
program operation, is then termed "garbage" because it cannot be subsequently
accessed or deleted.

A memory leak often occurs through the unintended reassignment of a pointer
to dynamically allocated memory:

```
char *pGarbage = new char( 'a' );
pGarbage = 0;                        // address to new memory lost
```

While the above statement only results in the loss of a single byte, if located in
a **for** loop, all memory available to the running program can be exhausted, as
in

```
for ( ; ; ; ) int *pGarbage = new char( 'a' );
```

which will typically slow down and then halt the computer, forcing a reboot.

12.13 Dangling pointers

While the assignment of a value to an uninitialized pointer is trapped by the
operating system, when values are written to and read from deallocated memory

the program will often continue to run normally. However, if this memory is later accessed by the program for another purpose such as a function call or a variable definition, or, in the case of dynamically allocated memory, for another running program, then unphysical results are produced or the program terminates unexpectedly. Such errors occur intermittently at unpredictable times widely separated from the execution of the incorrect source line, and are therefore often extremely difficult to identify in the absence of specialized software tools.

A dangling pointer is defined as a pointer to deallocated memory such as

```
int *pDangle = new int [20];
delete [ ] pDangle;      // the memory pDangle points to is deleted
```

Subsequently

```
pDangle[9] = 30;         // Bad error: uncertain result
```

places the value 30 into the unsafe memory location originally assigned to **pDangle[9]**. Subsequently dereferencing **pDangle**, however, still yields 30 unless the contents at this location were in the meantime overwritten by the program or operating system. As mentioned above, a recommended procedure for eliminating dangling pointers is to assign the **NULL** or **0** address to all pointers upon deallocation, as in

```
pDangle = NULL;
```

Attempting to read or write from the memory pointed to by **pDangle**

```
*pDangle = 30;           // Good error: program crash
```

now yields a runtime error since the operating system prohibits access to the null address.

A further dangling pointer error can occur if two pointers store the same memory location. If this memory is deallocated through one of these pointers whose address is set to **NULL**, the second pointer still accesses deallocated memory, e.g.

```
int *p = new int[20];
int *pDangle = p;
delete [ ] p;
p = 0;
pDangle[9] = 30; //Error: deallocated memory at pDangle's address!
```

A dangling pointer can also result from the deallocation of *compiler* allocated memory at the end of a block, as in

```
double *pDangle;
{
      double d = 3;
```

```
        pDangle = &d;
}                               // Error: d's memory deallocated
```

The above problem does not, however, occur with dynamically allocated memory, which persists until an explicit **delete** statement is issued:

```
double *pR;
{
        double *pD = new double[20];
        pR = pD;
}
pR[10] = 10;    // OK: pointer variable pD destroyed but not its memory
```

The memory associated with the pointer **pD** in fact cannot be automatically deallocated once **pD** has been destroyed, since a C++ runtime facility does not exist to determine whether additional pointers have been reassigned to point to **pD**'s memory as in the code above.

12.14 Pointers in function blocks

Since a function body also resides inside a block, dangling pointers can potentially arise if pointers are defined within a function, as in

```
void test( int *(&aPN), int *(&aPM ) {
        int k = 4;
        *aPN = k;           // OK: address stored in aPI unchanged
        aPM - &k;           // Error: will lead to dangling pointer
}                           // k's memory deallocated

main( ) {
        int n = 1, m = 2;
        int *pN = &n;
        int *pM = &m;
        test( pN, pM );
        cout << n << endl;    // Output: 4
        cout << *pM << endl; // Output: maybe 4 but unpredictable
}
```

The arguments **aPN** and **aPM** are *references* to pointers, and therefore correspond to alternative names for the pointers **pN** and **pM** defined in **main( )**. The value of the variable pointed to by **aPN** is changed inside **test( )**, so that **pN**, which points to the address space of the variable **n** in the main program, remains correctly allocated after the function call. On the other hand, the value of the pointer reference variable (i.e. the variable to which the pointer points) **aPM** is reassigned inside **test** to the memory address of the local **int** variable **k**. After the function block has terminated, **k** is destroyed and **pM** subsequently points to deallocated memory. If **aPM** were instead defined as a pointer argument, call-by-value semantics would apply and it would therefore contain a *copy* of the memory address in **pN**. Setting **aPM** to the address of **k** within the function block then does not affect the value of **pN**, which would therefore still point to the address of **n** defined in the main block rather than to deallocated memory.

12.15   Dynamic memory allocation within functions

Memory can be allocated to an external pointer from within a function by extending the procedure of the preceding example, namely

```
void allocate( int *(&aPI) ) {
    aPI = new int[10];
}                                    // OK: memory persists

main( ) {
    int *pI;
    allocate( pI );
    pI[5] = 3;
}
```

Although the compiler-allocated pointer variable **aPI** inside the function is destroyed when the function terminates, the memory assigned to this variable persists and therefore remains assigned to the parameter pointer variable, **pI**, in the calling program. The function argument must, however, again as in Section 12.14 be a *reference* to a pointer variable, otherwise **aPI** inside the function will be a *copy*, in a separate memory space, of the function parameter, **pI**, in the calling program. Setting the *value* of **aPI** inside the function to the location of the newly accessed memory then does not affect the *value* of **pI**, resulting in random memory access and a runtime error at the last line of the program. Recoding the above program but employing compiler-assigned memory yields a dangling pointer, since the lifetime of the compiler assigned memory then only extends to the end of the function block:

```
void allocateError(int *(&aPDangle) ) {
    int m[10];
    aPDangle = &m;
}                                    // Error: memory destroyed
```

Memory can also be assigned within a function and its address transferred to the calling program through a pointer return value:

```
float *pAssign( ) {
    float *pA = new float [20];
    return pA;
}                                    // OK: memory persists
```

However, this pointer again cannot point to compiler-assigned memory that is destroyed at the end of the function block:

```
float *pAssignDangle( ) {
    float a[20];
```

```
        float *pDangle = a;
        return pDangle;
}                                       // Error: memory destroyed
```

12.16 Dynamically allocated matrices

When matrices are dynamically allocated in scientific applications substandard procedures are often employed, degrading performance. This section accordingly discusses efficient allocation strategies.

Standard memory allocation. A dynamically allocated matrix (two-dimensional array) in C++ is implemented as a variable of type pointer to a pointer (see Section 12.9) that stores the address of an *array of dynamically allocated pointer variables*. Normally, the value of each pointer in this dynamically allocated array is assigned to the address of a further dynamically allocated array. That is, for an $N \times M$ matrix,

```
double **ppA = new double*[N];
for ( int loop = 0; loop < N; loop++ ) ppA[loop] = new double [M];
```

To deallocate the memory assigned to the matrix (but not **ppA**, the compiler-assigned pointer variable) the dynamically allocated array pointed to by each element of **ppA** must be deleted *before* the dynamically allocated pointer array assigned to **ppA** is deleted:

```
for ( int loop = 0; loop < N; loop++ ) delete [ ] ppA[loop];
delete [ ] ppA;
```

If just the second of these lines is present, the addresses to the arrays that store the actual matrix elements are (eventually) lost and the associated memory becomes garbage.

Optimal memory allocation. A significant drawback with the above dynamic allocation strategy is that memory for each matrix row is allocated through a separate call to **new**. Consequently, the operating system can assign widely separated areas in memory to successive rows. Numerous page faults can then result when iterating through all matrix elements. To insure contiguous memory, a one-dimensional array with $N \times M$ elements should first be allocated. The values of the pointers of a dynamically allocated array of N pointers are then set to the address of every Mth element of the array. For a 2×3 matrix of doubles,

```
double *pB = new double[6];               // contiguous
double **ppA = new double*[2];
for ( int loop = 0; loop < 2; loop++ ) ppA[loop] = pB + 3 * loop;
```

To deallocate the dynamically assigned memory, the memory associated with the underlying array **pB** should be deleted after the pointer array **ppA**:

```
delete [ ] ppA;
delete [ ] pB;
```

If **pB** is deallocated first, the matrix **ppA** will point to unallocated memory, potentially leading to incorrect memory access if the matrix is mistakenly employed before deletion.

12.17   Dynamically allocated matrices as function arguments

A dynamically allocated matrix can be passed to a function through an argument of pointer-to-pointer type:

```
void print ( int **aPPA, int aN, int aM ) {
    for ( int outerLoop = 0; outerLoop < aN; outerLoop ++ ) {
        for ( int innerLoop = 0; innerLoop < aM; innerLoop ++ )
            cout << aPPA[innerLoop][outerLoop] << ' ';
        cout << endl;
    }
}

main ( ) {
    int **ppA = new int *[2];
    ppA[0] = new int[2];
    ppA[1] = new int[2];
    ppA[0][0] = ppA[0][1] = ppA[1][0] = ppA[1][1] = 6;
    print ( ppA, 2, 2 );                    // Output: 6 6
}                                           //               6 6
```

However, the above print function *cannot* be used for a statically allocated matrix,

```
int B[2][2] = { 1, 2, 3, 4 };
print ( B, 2, 2 );                          // Error!
```

since, in this case, the compiler requires that the second matrix dimension be specified in the function parameter list (e.g. **void print(double *A[2], int aN, int aM)**) because the type of **B** is that of an *array of two-component integer arrays*.

Similarly, a dynamically but not a statically allocated two-dimensional array can be assigned to a variable of type pointer to a pointer:

```
int A[2][2] = { 1, 2, 3, 4 };
int **ppB = new int*[2];
ppB[0] = new int[2];
ppB[1] = new int[2];
int **ppD = B;                              // OK
int (*ppAE)[2] = A;                         // OK
ppD = A;                                    // Error!
```

The manipulation of complex types, especially in conjunction with dynamic memory allocation, requires a detailed understanding of their underlying structure.

12.18 Pointer data structures and linked lists

Dynamic memory allocation offers a greater range of methods for manipulating data than static allocation since the size of the structure into which the data are placed can expand or contract as elements are added or deleted. Further, through pointers the structure properties can reflect the relationship of different data elements. As an example, a linked list is a collection of data such that each element in the list contains both data and the address of the succeeding element in the list. Consequently, data can be read sequentially from the initial data element to the final data element. Functions that print, insert new data at a given position in the list, or delete an element from the list accordingly process the data by iterating through the list elements. The non-zero elements and their indices of e.g. a sparse matrix can be efficiently stored in such a list.

A single list component is termed a node. A node that stores a single data value together with a pointer to an address of a second node is given by

```
class Node {
public:
        Node *iPNext;
        double iValue;
};
```

A pointer of type **Node*** containing the address of the first list element constitutes the "head" of a linked list, while **iPNext** in the last element or tail of the list is set to **0** or **NULL**. The latter choice enables a logical test for the end of the list in the same manner as a zero character terminates a string. Hence, a three-element list can be generated by

```
Node NLast = { 0, 3 };
Node NMiddle = { &NLast, 1 };
Node NFirst = { &NMiddle, 8 };
Node *pHead = &NFirst;
```

Exploiting the presence of the null pointer at the tail, a **for** loop can iterate through the list items as in the following **printList( )** function that writes out the data in the list:

```
void printList( Node * const aPHead ) {
        for ( Node *pN = aPHead; pN != NULL; pN = pN -> iPNext )
                cout << ( pN -> iValue ) << endl;
}
```

Since a **Node** element is reached by navigating through each intervening element starting from the head, variations of the above construct are required in order to achieve more complicated functionality. For example, to add a **Node** element to the end of a non-empty list, the elements must be iterated over until the **NULL** pointer is detected. The **next** pointer of this final element is then reassigned to the new node, whose **next** pointer is set to **0**:

```
void addNodeAsLast( Node *aPHead, Node *aPNew ) {
     Node *pN;
     if ( aPHead == 0 ) exit( 0 );
     for ( pN = aPHead; pN -> iPNext; pN = pN -> iPNext );
     aPNew -> iPNext = 0;
     pN -> iPNext = aPNew;
}
```

Our list example can then be rewritten as

```
main( ) {
     Node NFirst = { 0, 8 };
     Node NLast = { 0, 3 };
     Node NMiddle = { 0, 1 };
     Node *pHead = &NFirst;
     addNodeAsLast( pHead, &NMiddle );
     addNodeAsLast( pHead, &NLast );
     printList( pHead );
}
```

A **List** class combines the **Node** class with functions that manipulate lists while employing dynamic memory allocation to generate and remove nodes. A bidirectional list that can be traversed in either the forward or the reverse direction is generated by including a second pointer, **iPPrevious**, in the **Node** class that points to the preceeding node. A binary tree is then obtained if each of the pointers **iPPrevious** and **iPNext** of a given node points either to **NULL** or to a separate node that is not pointed to by any other node. That is, each node may point to zero, one or two other nodes, but is pointed to by just a single node. Operations on tree structures are again implemented through recursion and are initially somewhat difficult to program. In general, for dealing with physical problems that are naturally formulated in terms of non-trival data structures one should employ an appropriate collection class package.

Chapter 13
Memory management

The following two chapters consider advanced or infrequently used features of C++ programming. We first examine the dynamic allocation of memory within objects, as occurs when memory for an internal member variable of pointer type is dynamically allocated in a class constructor. If the resulting object is deallocated, its internal variables, including the pointer, are destroyed. However, the dynamically assigned memory persists, generating a memory leak, unless a special "destructor" function is introduced. Similar problems occur when such objects are copied to or initialized by a second object.

13.1 The this pointer

Memory management requires additional C++ constructs, the first being the **this** pointer which exists in all (non-static) classes and points to the calling object. *Dereferencing the **this** pointer generates the calling object itself* so that in the program

```
class A {
    public:
    int iA;
    void print ( ) { cout << (this -> iA) << endl; }
};
```

this -> iA is interchangeable with **iA** since ***this** represents the object that calls the **print( )** function.

A fundamental application of the **this** pointer is given by

```
class C {
    public:
    int iM;
    C& print ( ) {
        cout << iM << " " " ";
        return *this;
    }
    C* pAdd( int aN ) {
        iM += aN;
```

```
            return this;
        }
};

main ( ) {
    C C1 = { 2 };
    C1.pAdd( 3 ) -> print( ).pAdd( 3 ) -> print( ); //Output: 5 8
}
```

Since the . and -> functions are left-associative, the last line of the **main( )** program is interpreted as **(((C1.pAdd(3)) -> print( )).pAdd(3)) -> print( );**. Consequently the **pAdd( )** function of **C1** is applied first, returning a pointer to the calling object, namely **C1** itself. Next, this pointer is dereferenced through the pointer-to-member operator and the **print( )** function is called, which returns a reference to **C1**. Thus the combined effect of these operations is to return the initial object **C1** with modified internal data members, to which a further series of add and print functions is applied. While the above program still functions if **C** is employed in place of **C&** as the return type of the **print( )** function, a superfluous and potentially problematic (see Section 13.6) copy operation occurs when the **C** object is returned.

13.2   The friend keyword

If a class, **A**, designates a particular function or class external to itself as a **friend**, the named construct can access the private members of **A** directly without being restricted to calling the member functions of **A**'s public (or possibly protected) interface. That is, in

```
class A {
    int iJ;
    friend class B;
};
```

all the private data and member functions of **A** appear as public members in class **B** (which can be a subclass of **A**) so that the following code is valid:

```
class B {
    A* iPA;
    public:
    void printA( ) { cout << iPA -> iJ << endl; }
    B( int aJ ) {
        iPA = new A;
        iPA -> iJ = aJ;
    }
};

main ( ) {
    B B1( 5 );
    B1.printA( );                              // Output: 5
}
```

To grant a single function **int bfun(int bI)** in class **B** access to the private members of **A**, a line

```
friend int B::bfun( int );
```

should be included in the (public or private) interface of class **A**. Since the friend function, **bfun**, is not a member of the class **A**, it does not have an associated **this** pointer to **A**. Although often convenient, the **friend** construct violates the object-oriented principles of encapsulation and information hiding.

13.3 Operators

C++ operators can be redefined or extended to user-defined classes through overloading. However, although any C++ operator, with the exception of the scope operator, ::, the dereferencing operator, *, the member-of operator, ., the **sizeof** operator and the if-then-else ternary operator, ?:, can be overloaded, new operator symbols cannot be introduced and existing precedence rules cannot be altered. Accordingly, certain operators have illogical precedence. For example, the C++ stream extraction and insertion operators overloaded preexisting high-precedence C operators for bit insertion and extraction, invalidating seemingly meaningful constructs such as **cout << i = 3 << endl;**. Two procedures for overloading operators exist and are discussed individually below.

Overloading through friend functions. An operator acting on class members can be introduced through a friend function that accesses the internal class variables. To illustrate, the class **Vector** below possesses two private data members, **iX** and **iY**. The binary addition operator + is overloaded such that adding two **Vector** objects **Vector1** and **Vector2** with internal data members **iX1**, **iY1** and **iX2**, **iY2** yields a new **Vector** object with data members **iX1 + iX2** and **iY1 + iY2**. When expressed as a **friend** function, the + operator possesses the special signature **operator+ (V1, V2)**, in which the first and second arguments are the expressions to the left and to the right of the + sign, respectively. This yields the class definition

```
class Vector {
public:
      double iX, iY;
      friend Vector operator+ ( Vector, Vector );
      Vector( double aX, double aY ) : iX( aX ), iY( aY ) { }
      void printVector( ) { cout << iX << "\t" << iY << endl; }
};
```

The **friend** function is the two-argument global function

```
Vector operator+ ( Vector aV1, Vector aV2 ) {
      Vector Temp( 0, 0 );
      Temp.iX = aV1.iX + aV2.iX;
      Temp.iY = aV1.iY + aV2.iY;
      return Temp;
}
```

Subsequently the operator can be called, as in

```
main ( ) {
        Vector V1( 1, 2 ), V2( 2, 3 );
        Vector V3( 0, 0 );
        V3 = V1 + V2;
        V3.printVector( );                              // Output: 3 5
}
```

To prevent copying of the input arguments, the **operator** function signature can be replaced by **friend Vector operator+ (const Vector&, const Vector&);**. The statement **V3 = V1 + V2;** in **main( )** can even be replaced with the equivalent **V3 = operator+ (V1, V2);**.

Inclusion in the class definition. The second technique for operator overloading incorporates the operator directly into the class definition. The operation **Vector1 + Vector2** is then interpreted as **Vector1.operator+(Vector2)**, where + is a member function of the **Vector** class and **Vector2** is an argument of type **Vector**. Since the member variables **iX** and **iY** of **Vector1** are directly accessible, the code takes the form

```
class Vector {
        public:
        double iX, iY;
        Vector operator+ ( Vector );
        Vector( double aX, double aY ) : iX( aX ), iY( aY ) { }
        void printVector( ) { cout << iX << "\t" << iY << endl; }
};

Vector Vector::operator+ ( Vector aV ) {
        Vector Temp( 0, 0 );
        Temp.iX = iX + aV.iX;
        Temp.iY = iY + aV.iY;
        return Temp;
}
```

Observe that here the *binary* operator + possesses only a *single* argument and can therefore be called either through the standard notation **V3 = V1 + V2;** or the alternate notation **V3 = V1.operator+ (V2);**. However, the **friend** implementation of binary operators is generally preferable, since both operator arguments are handled symmetrically with respect to implicit or user-defined conversions.

The extraction operator is often overloaded at global scope to output the internal state of an object in a convenient format, as, for example,

```
ostream& operator<< ( ostream &out, Vector V ) {
        out << "The vector components are"
            << V.iX << " and " << V.iY << endl;
        return out;
}
```

The statement **cout << V3;** then yields **The vector components are 3 and 5**. If the internal class variables **iX** and **iY** are **private**, the operator must be a **friend** of the **Vector** class.

13.4 Destructors

Having introduced operators and friend functions, we now discuss memory management in classes that *dynamically allocate memory within constructors*, as in

```
class A {
   public:
   A( int aValue = 0, int aSize = 2 ):iValue( aValue ), iSize( aSize )
   {
        iPArray = new int [aSize];
   }
   void print ( )
   {
        cout << iValue << endl;
   }
   int iSize;
   int *iPArray;
   int iValue;
};
```

This class definition will in general create memory leaks unless *three* auxiliary functions, the destructor, assignment operator and copy constructor, are introduced.

A destructor is called when an object of type **A** goes out of scope. Without a destructor, the compiler-assigned internal variables of **A1**, including the pointer **iPArray**, are automatically deallocated. However, the memory that the **iPArray** variable points to is dynamically assigned through the **new** statement in the constructor of **A1** and is therefore not deallocated, yielding a memory leak. Of course, since the internal class variables in **A** are public, this could be rectified by

```
{
   A A1( aValue );
   delete [ ] A1.iPArray;
}
```

However, if **iPArray** were declared **private**, a separate **public** member function would have to be introduced into the **A** class definition to enable memory deallocation in the above manner. To automate this procedure a "destructor" member function is invisibly called whenever an object of the class type is destroyed in the same manner as a constructor is called when an object is created. Since the destructor cannot be called explicitly and does not return a variable, it does not possess an independent name or return value. Instead, the destructor is denoted in the class definition by the class name preceded by the ~ character (the symbol not followed by the class name is intended as a mnemonic for object destruction). In the above example, an appropriate destructor is

```
~A( ) {
      delete [ ] iPArray;
}
```

If a *pointer* to a dynamically allocated vector object is destroyed, only the compiler-allocated pointer variable is deallocated, while the dynamically allocated memory resident at the value of the pointer persists. Therefore *an explicit delete statement must be supplied* as in (A's constructor is a default constructor since default values are provided for all arguments)

```
{
      A *pA1 = new A;
      delete pA1;                        // Calls pA1's destructor
}
```

Deleting the memory of the object assigned to the pointer **pA1** calls its destructor. If memory is not always dynamically assigned to a given pointer by the constructor, since **delete** or **delete [ ]** operators can always safely be applied to null pointers, any pointer that will be the target of a **delete** statement in the destructor should be set to the null pointer in the constructor when dynamic memory is not allocated.

Before an object is constructed, its base classes are first constructed and initialized in the order in which they are declared. Within each class, class members are again initialized in order of declaration. The destructor destroys memory in the opposite order to that in which it is allocated by the constructor.

13.5 Assignment operators

Recall that, when an object is assigned to a second non-**const** object of the same type, the values of all compiler-allocated internal variables of the first object are copied to the corresponding variables of the second object. However, memory leaks result if memory is dynamically allocated to an internal pointer variable through the constructor, as in

```
main ( ) {
      A A1( 3 );
      A AGarbage;
      AGarbage = A1;
      AGarbage.print( );                 // Output: 3
}
```

In the line **AGarbage = A1**, the *values* of all *compiler*-assigned variables, namely **iPArray**, **iSize** and **iValue**, are copied from **A1** to **AGarbage** in the **main( )** function, a procedure termed a "*shallow copy*". Therefore, in the **main( )** program, the variable **iPArray** in **AGarbage** subsequently points to the same memory as **iPArray** in **A1**. While the program then functions properly (except that a change to the array at **iPArray** in either **A1** or **AGarbage** will propagate to the other object), a memory leak has occurred. In particular, **AGarbage** when created through the default constructor dynamically allocates memory for an array of two integers and the address of this memory is stored in its **iPArray** pointer. When the *value* of (the memory address stored in) **AGarbage.iPArray** is overwritten by **A1.iPArray** through the assignment statement **AGarbage = A1;**, the memory address of the original dynamically allocated array in **AGarbage** is lost and cannot subsequently be reclaimed by the operating system until program termination, even if a destructor function is specified.

A related but more severe error occurs in

```
A ADangle(2);
{
     A A1(3);
     ADangle = A1;
}                // A1 is destroyed: ADangle.iPArray is deallocated.
ADangle.iPArray[0] = 1;//Intermittent error: illegal memory access
```

Since the memory locations pointed to by **iPArray** in both **A1** and **ADangle** are identical after the assignment statement, once **A1** has been destroyed by exiting the block in which it is defined, **ADangle.iPArray** points to deallocated memory.

The above difficulties are resolved by overloading the assignment operator for objects of type **A** to prevent the value of the pointer variable, **iPArray**, in **A1** from being copied to the pointer variable in **A2** in a statement **A2 = A1;**. Instead, if the length of the dynamically allocated array, **A2.iPArray**, is greater than or equal to the length of **A1.iPArray**, each *element* of **A1** must be copied into the corresponding element of **A2** (a "*deep copy*"). If the size of the array in **A2**, on the other hand, is less than the size of **A1**, the dynamically allocated memory in **A2** should be first deleted and a new array allocated with the same dimension as **A1.iPArray**, followed by a deep element-by-element copy operation. Also, the user-defined assignment operator must copy all compiler-assigned variables from **A1** to **A2**, since the default assignment operator that normally performs this operation has been replaced. The signature of an assignment operator is **typeName& operator = (const typeName&)**, where the **const** keyword is optional. Accordingly, for our sample class, the overloaded assignment operator takes the form

```
class A {
// ... code for internal variables, constructor, destructor, etc.
     A& operator = ( const A& aA ) {
// Ensure that the object is not copied onto itself
          if ( this == &aA ) return *this;
// Copy compiler-assigned variables
```

```
          iValue = aA.iValue;
          iSize = aA.iSize;
// Assign new memory to the pointer variable if the
// dimension of the preexisting array is smaller
// than that required by the new array.
          if ( iSize < aA.iSize ) {
                    delete[ ] iPArray;
                    iPArray = new int[aA.iSize];
          }
// Perform a deep copy of the dynamically allocated elements
// in aA to those of the calling object.
          for ( int loop = 0; loop < aA.iSize; loop++ )
                    iPArray[loop] = aA.iPArray[loop];
          return *this;
     }
};
```

The first line of the above function prevents the object's memory from being released in the case that the object is copied onto itself. A simpler procedure for transferring the elements of the array **aA.iPArray** to **iPArray** is provided by the **memcpy** function (the last argument of this function is the number of bytes to copy):

```
memcpy( iPArray, aA.iPArray, iSize * sizeof( int ) );
```

The **memcopy** function resembles **strcpy(s2, s1)**, which copies a string **s1** into a second string **s2**, and can duplicate arrays of any dimension.

13.6 Copy constructors

Finally, when an object is defined and simultaneously initialized with a second object of the same type, a *copy constructor* as opposed to the assignment operator is invoked. This occurs in four cases. The first two of these are

```
A A1;
```

and

```
{
     A A2Garbage( A1 );
     A A3Garbage = A1;
}
```

which are implemented identically, although the equality sign in the declaration of **A3Garbage** incorrectly implies that the assignment operator is invoked. The last two arise when an object of type **A** is passed to or returned by a value from a function, both of which occur when **test( )** is called in the examples below:

```
A test( A aAGarbage ) {
     aAGarbage.print( );
```

```
        return aAGarbage;
}
```

and

```
main( ) {
        A A1( 1 );
        A A2Garbage;
        A2Garbage = test( A1 );
}
```

In each of the above cases, compiler-assigned variables are copied from one **A** object to a second **A** object through a shallow copy and the address of the dynamically assigned memory in the second object is again lost. However, these copies do *not* proceed through the assignment operator but rather through the default *copy constructor*, which possesses the signature **classType (const classType&)**. A copy constructor that performs a deep copy is implemented almost identically to the overloaded assignment operator:

```
A ( const A& aA ) {
// Copy compiler-assigned variables.
        iValue = aA.iValue;
        iSize = aA.iSize;
// Assign new array memory to the pointer member variable.
        iPArray = new int[aA.iSize];
// Perform a deep copy of the dynamically allocated elements in aA
// to those of the calling object.
        memcpy( iPArray, aA.iPArray, iSize * sizeof(int) );
}
```

The reference parameter in the function signature is required since C++ does not allow a constructor in a class to have a non-reference object of its own class type as a parameter. A copy constructor can also be coded by invoking the overloaded assignment operator:

```
A( const A& aA ) {
        iPArray = new int[aA.Size];
        *this = aA;
}
```

To prohibit copying of objects of a given class, a private copy constructor without a body can be introduced:

```
A ( const A & ) { };
```

Chapter 14
The static keyword, multiple and virtual inheritance, templates and the STL

This final C++ programming chapter summarizes several additional features of the C++ language, including the **static** keyword, unions, bit fields, virtual and multiple inheritance, templates and the standard template library (STL).

14.1   Static variables

*Adding the **static** keyword to a variable's definition statement extends its lifetime from the beginning to the end of the program*, although it is not visible outside the block in which it is defined unless it can be accessed through the scope-resolution operator. Thus *a **static** variable defined within a function retains its value from one function call to the next, but cannot be accessed outside the function block* (the body). Any **static** variable that is defined outside the body of a class is automatically initialized to zero in the absence of an initializer. A **static** variable that records the number of times a function is called is coded as

```
void f( ) {
      static int iStatic = 1;
      cout << iStatic++ << endl;
}

main( ) {
      f( );
      f( );
}                               // Output is 1 then 2.
```

The **static** keyword often simplifies programming but violates the principle of encapsulation.

14.2   Static class members

*If a member variable or function is declared **static** within a class definition, a single instance of this class member again exists over the entire lifetime of the program, irrespective of whether any actual objects of the class are ever instantiated.* However, a static variable instance can be accessed outside the class since it has an associated scope identifier. Consequently, *if a class has a **static** member,*

its value is the same for all objects of the class. Since **static** members are independent of the state or even existence of the objects, *they can be accessed through
their name preceded by their class name followed by the scope-resolution operator as well as through the normal member-of or pointer-to-member operators.*
Class member variables, whether **public**, **protected** or **private**, that are declared
static *must* have a corresponding initialization statement at global scope (outside
the class definition) that does *not* include the **static** keyword, since these variables
exist even if no objects of the class are instantiated and therefore cannot be initialized within a constructor. If a value is not specified in the initialization statement,
the variable is automatically initialized to zero as in the following example:

```
class C {
  public:
  C( ) { cout << ++iStatic << " ";}
  private:
  static int iStatic;
};

int C::iStatic;                        // Zero by default

main ( ) {
  C C1;
  C C2;
}                                      // Output: 1 2
```

A **static** member function, like a **static** variable, exists throughout program
execution regardless of the number or existence of the class objects. Therefore,
such a function can only change static class variables (as illustrated above, non-
static functions can change **static** data as well). Further, a **static** member function
does not have a **this** pointer, because it is not associated with any particular object.
An example of **static** member variables together with functions that employ or
change their values is given below:

```
class C {
public:
  static int iStatic;
  int iVariable;
  C ( int aVariable ) : iVariable( aVariable ) { }
  static void setStatic( int aStatic ) { iStatic = aStatic; }
  static int calculate( C& aC ) { return iStatic * aC.iVariable; }
};

int C::iStatic = 2;                    // Zero by default; here 2

main( ) {
  C::setStatic( 3 );
  C::iStatic( 4 );
  C C1( 2 );
  C1.iStatic = 5;
  cout << C::calculate( C1 ) << endl;   // Output: 10
}
```

(Note that, since **calculate( )** does not depend on any particular **C** object, it can be declared **static**.)

14.3   Virtual functions

Recall that a derived class object can be employed anywhere in a C++ program that a base class object is expected. That is, if a class, such as **Student**, is derived from a second class, e.g. **Person**, a **Student** represents a specialized form of a **Person**. Therefore, any behavior of a **Person** should also be applicable to a **Student**. *However, if a derived class overrides one or more of the functions (methods) of the base class, the programmer can in certain cases specify whether the derived class or the base class definition should be employed when a derived class object is substituted for a base class object. This requires, however, pointer or reference variables, since storing the address of a derived class object in a base class pointer does not affect the memory assigned to the object.* In contrast, assigning a derived class object to a base class object truncates the memory of the derived class variable through a conversion operation that discards the specialized features of the derived class.

The type of a pointer or reference is normally associated at compile time with the type specified in its definition statement. Accordingly,

```
class Person{
      public:
      void print( ) { cout << "Person " << endl; }
};
class Student : public Person {
      public:
      void print( ) { cout << "Student " << endl; }
};

main( ) {
      Student Student1;
      Person* pPointerToStudent1 = &Student1;
      Person& RefToStudent1 = Student1;
      pPointerToStudent1 -> print( );
      RefToStudent1.print( );
      Person Person2 = Student1;
      Person2.print( );
}
```

yields the output **Person Person Person** since, although the pointer and reference variables refer to **Student** objects, they are declared **Person*** and **Person&**, respectively. However, *this behavior can be altered for any function that appears both in the base class and in the derived class by including the keyword* **virtual** *in its base class definition* (including the **virtual** keyword also in the derived class, while unnecessary, is still recommended for clarity). If a derived class pointer or object is then assigned to a base class pointer or reference variable, the derived class object properties are instead employed at *runtime*. Hence, if different base or derived class objects are stored in a base class pointer in response to e.g. user input and subsequently accessed, different behaviors result. To illustrate, replacing the **Person** class above by the code

```
class Person{
     public:
     virtual void print( ) { cout << "Person " << endl; }
};
```

yields **Student Student Person** when the modified program is executed.

A function that is virtual in a base class remains virtual in all derived classes, including those that are derived through multiple layers of inheritance. If a virtual function is not defined in a particular derived class, the definition in the parent class that is closest in inheritance level to the derived class is employed. *While constructors cannot be virtual, destructors should be declared virtual in the base class so that, if a derived object is deallocated through a pointer to a base type at runtime, the destructor of the derived type is called.*

14.4 Heterogeneous object collections and runtime type identification

Virtual functions are particularly convenient for processing groups of objects that are derived from the same base class. *Since every member of the derived class is also a member of the base class, these can be stored as a "heterogeneous" collection (e.g. a data structure such as an array, list or queue) of pointers or references to base class objects. If a function is declared* **virtual** *in the base class, its behavior when accessed through the collection will be that associated with its object type.* Since the object type is resolved at runtime, the objects can further be placed into the collection at runtime according to user selections or logical outcomes. For example, assuming that **print( )** is declared **virtual** in the **Person** base class, a two-element heterogeneous object collection can be created and accessed in our previous example as follows:

```
main( ) {
  Person **ppArray = new Person*[2];
  Person P1;
  Student S1;
  int select;
  for ( int loop = 0; loop < 2; loop++ ) {
    cout << "Insert 0 to create a Person, 1 to create a Student ";
    cin >> select;          // Sample input: 0 1
    cout << endl;
    if ( !select ) ppArray[loop] = &P1;
    else ppArray[loop] = &S1;
  }
  for ( int loop = 0; loop < 2; loop++ ) {
```

```
    ppArray[loop] -> print( );     // Sample output: Person Student
// cout << endl << typeid( *ppArray[loop] ).name( ) << endl;
// Student *pS2 = dynamic_cast<Student *> ( ppArray[loop] );
// if ( pS2 ) cout << ( typeid( pS2 ) ==
//     typeid( Student * ) ) << endl;
  }
}
```

The **dynamic_cast<typename>** operator above *downcasts* the base class pointer to a derived class pointer. If the conversion fails, the operator returns the null pointer of the type specified by **typename**. Further, the pointer type can be established during program execution through the **typeid( )** function defined in the **<typeinfo>** header file. This function, which can be applied to pointers, references and dereferenced pointers, generates an object of the **type_info** class. The **name( )** member function of this class returns the type of its argument as a string. The procedure is illustrated through the commented lines in the program above, which print out the type of the dereferenced pointers stored in **ppArray**, namely **Person** or **Student** (in Dev-C++ preceded by a number) followed by **1** for each value of **loop** such that **ppArray[loop]** is a pointer to a **Student**.

14.5 Abstract base classes and interfaces

Since a pointer or reference to a virtual function associates its implementation with the type of the containing object, if base class objects are never constructed, the bodies of one or more virtual functions are superfluous and can be omitted from the base class. Such a class is then termed an abstract (base) class. A class that derives from an abstract base class must supply bodies for all missing functions in order to construct objects of its type. The base class thus provides an interface to which physical derived classes must conform by supplying all requested function definitions. A function declaration (prototype) appearing in a base class without a corresponding function definition (body) is termed a pure virtual function, while a class that contains one or more pure virtual functions is labeled an abstract class.

Abstract classes are often employed as base classes for heterogeneous object collections since derived class elements can always be placed into arrays of pointers to base class objects:

```
#include <iostream.h>

class Person{
      public:
      virtual void print( ) = 0;
};

class Student : public Person {
      public:
      void print( ) { cout << "Student" << endl; }
};
```

```
class Worker : public Person {
       public:
       void print( ) { cout << "Worker" << endl; }
};

main( ) {
       Student Student1;
       Worker Worker1;
       Person *PArray[2] = { &Student1, &Worker1 };
       PArray[1] -> print( );                    //  Output: Worker
}
```

Abstract base classes provide an *interface* that derived classes must conform to, as opposed to standard inheritance, which instead provides an *implementation* that is adopted by the derived classes. In interface inheritance, the compiler verifies that the derived classes provide all required behaviors.

14.6 Multiple inheritance

A derived class may inherit the attributes (internal variables and functions) of any number of parent classes. As a simple illustration of a class **C** that inherits the member variables and functions of two base classes, **A** and **B**,

```
class A {
       protected:
       int iA;
       A( int aA ) : iA( aA ) { }
       void print( ) { cout << iA << endl; }
};
class B {
       protected:
       int iB;
       B( int aB ) : iB( aB ) { }
       void print( ) { cout << iB << endl; }
};
class C: public A, public B {
       public:
       int iC;
       void print( ){ B::print( ); }
       C( int aA, int aB, int aC ) : A( aA ), B( aB ) { iC = aC; }
};
```

Since the constructor of class **C** passes arguments to the base class constructors of **A** and **B** through its initialization list, these are invoked before **C** is constructed. However, since **iA** and **iB** are inherited member variables of **C**, its constructor can also be written as

```
C( aA, aB, aC ) : iA( aA ), iB( aB ), iC( aC ) { }
```

Note that the **print( )** functions of the parent classes of **C** are distinguished in class **C** through the scope-resolution operator.

14.7   Virtual inheritance

An ambiguity in multiple inheritance occurs if e.g. two classes **Student** and **Worker** both inherit from a common base class **Person** while a further class **StudentEmployee** inherits from both **Student** and **Worker**. In this case, a data member or function such as **print( )** in the **Person** class is inherited in **StudentEmployee** through both **Student** and **Worker**. Thus, **print( )** is inherited twice (whether or not it is declared **virtual** in the base class or overridden in one or both of the **Student** or **Worker** classes), once through **Student** and once through **Worker**. As a result, calling **print( )** in **StudentEmployee** requires a scope-resolution operator such as **Student::print( )**, since the compiler cannot resolve which of the two inherited **print( )** statements is intended (even if they are identical). Secondly, a pointer or reference to **StudentEmployee** cannot be employed where a pointer or reference to a **Person** is required, as evidenced by

```
class Person{
      public:
      void print( ) { cout << "person" << endl; }
};

class Student : public Person {
};
class Worker : public Person {
};
class StudentEmployee : public Student, public Worker {
};

main( ) {
      StudentEmployee SE1;
      SE1.print( );          // Error: ambiguous member
      Person *pP1 = &SE1;    // Error: cannot convert to Person*
      pP1 -> print( );
}
```

which yields the error messages shown. To insure that only a single copy of the base class and hence **print( )** is inherited by **StudentEmployee**, the **Student** and **Worker** class signatures must be replaced with

```
class Student : virtual public Person {
};
class Worker : virtual public Person {
};
```

14.8   User-defined conversions

While many promotions and conversions among built-in types such as those from **int** to **float** or **double** are implicit in C++, additional conversions are programmed either as conversion operators in class definitions or as single-argument constructors. That is, if a class **A** is to be automatically converted to a class **B**, a conversion operator can be introduced in the **A** class with the special signature **operator B( )**; that implements a cast from **A** to **B**. Alternatively, a single-argument constructor can be defined with the signature **B(A)** in the **B** class. To illustrate both procedures, the following converts a **Fahrenheit** to a **Celsius** temperature according to $°F = 9/5 \, °C + 32$ through a single-argument constructor and converts a **Celsius** temperature to a **double**:

```
class Celsius;

class Fahrenheit {
    public:
        double iDegrees;
        Fahrenheit( Celsius aCelsius );
        Fahrenheit( double aDegrees ) : iDegrees( aDegrees ) { }
        operator double( ) { return iDegrees; }
};

class Celsius {
    public:
        double iDegrees;
        Celsius ( double aDegrees ) { iDegrees = aDegrees; }
        Celsius( Fahrenheit aFahrenheit ) {
           iDegrees = ( aFahrenheit.iDegrees - 32.0 ) * 5.0 / 9.0; }
        operator double( ) { return iDegrees; }
};

Fahrenheit::Fahrenheit( Celsius aCelsius ) {
    iDegrees = aCelsius.iDegrees * 9 / 5 + 32.0; }
void printState( Celsius aC ) { aC <= 0. ? cout <<"Below Freezing"
    << endl : cout << "Above Freezing" << endl; }

main ( ) {
    Fahrenheit F( 34.0 );
    Celsius C = F;
    cout << double( C ) << endl;     // Output: 1.11111
    printState( F );                 // Output: above freezing
}
```

To prohibit single-argument constructors from automatically functioning as conversion operators, the keyword **explicit** can be included in their type definitions.

14.9 Function templates

A class or function can be applied to variables or objects of several different types without coding many nearly identical copies through the **template** keyword. This keyword transforms class names in the program into *metaparameters* that can be assigned different values. *At compile time, a separate copy of the template component is generated by the compiler for each distinct set of metaparameter types.* A function template (which is not universally supported) is illustrated by the following example, which copies arrays of any specified class type after adding 32, the offset between large and small characters in the ASCII character set:

```
template <class C1, class C2> void copy( C1 aOutput[ ],
    C2 aInput [ ], int aN ) {
    for( int loop = 0; loop < aN; loop++ )
        aOutput[loop] = aInput[loop] + 32;
}
```

The keywords **class** and **typename** are interchangeable in the template argument list. The above function can be called without template parameters as in the first call to **copy( )** in the program below. The class identifiers **C1** and **C2** are then automatically determined from the types of the function parameters so that **C1** and **C2** evaluate to **double** and **int**, respectively. Alternatively, as in the next line of the program, the template parameters can be explicitly supplied in the parameter list. Finally, in the third call to **copy( )**, the template function arguments are implemented as **char** arrays:

```
main( ){
    double myDouble[10], myDoubleNew[10];
    int myInt[10] = { 1 };
    copy ( myDouble, myInt, 10 );
    cout << myDouble[0] << endl ;          //Output : 33
    copy <double, double> ( myDoubleNew, myDouble, 10 );
    char c1[5], c2[5] = { 'A' };
    copy ( c1, c2, 5 );
    cout << c1[0] << endl;                 //Output: a
}
```

Default template parameter values cannot be specified in a function template; however, a non-template specialization of a function can overload a template function with the same name. Hence, if a second non-template **copy( )** function were employed in the above program with specific parameter types such as

```
void copy( char aOutput[ ], char aInput [ ], int aN ) {
    for ( int loop = 0; loop < aN; loop++ )
        aOutput[loop] = aInput[loop] + 32;
}
```

this would be called in place of the template function whenever the **copy** function is passed two character arrays, as in the third function call in **main( )** above.

14.10 Templates and classes

A **template** class is coded analogously to a function template. An example, which requires that the class parameters that are substituted for C1 and C2 possess appropriately overloaded stream insertion operators (<<), is given by

```
#include <sstream>

template <class C1, class C2 = int, int aN = 1> class Logical {
        C1 iC1;
        C2 iC2;
        int iN;
public:
        Logical ( C1 aC1, C2 aC2, int aN ) : iN( aN )
            { iC1 = aC1, iC2 = aC2; }
        void and( ) { cout << ( iC1 && iC2 ) << " " << iN << endl; }
        string asString( );
};

template <class C1, class C2, int aN>
          string Logical<C1, C2, aN>::asString( ) {
        stringstream sout;
        sout << iC1 << ' ' << iC2 << endl;
        return sout.str( );
}
```

Then the output of

```
main( ) {
Logical<int, bool> Test( 0, 0, 3 );
Test.and( );
cout << Test.asString( );
}
```

is

```
0 3
0 0
```

Note the placement of the scope operator in the **asString( )** function definition. The "non-type" template parameter **aN** cannot be a floating-point variable, class, pointer or array and must be assigned a compile-time constant of a compatible type (for example, if **aN** is of type **bool** it can be set to either of the **bool** constant values **true** or **false** or to a variable of type **const bool**). If default parameters are omitted from the template parameter list, their default values are employed. As in the case of functions, default parameters must appear last in the template argument list.

A static member variable of a templated class possesses a separate realization for each template instantiation, i.e. each time the template is called with a distinct set of template parameters. Thus

```
template< class T > struct MyStatic { static int iI; };
template< class T > MyStatic < bool >::iI;
template< class T > MyStatic < char >::iI;

main( ) {
        MyStatic< bool > MB1, MB2;
        MyStatic< char > MC1, MC2;
        MB1.iI = 2;
        MyStatic< char >::iI = 'a';
        cout << MB2.iI << ' ' << MyStatic<bool>::iI
            << ' ' << MC2.iI << endl;
}
```

yields the output 2 2 97. Referring to static template variables through the syntax **classname<typename>::staticvariablename** identifies the variable as static and is therefore recommended.

Template arguments can encompass further template parameters, constant expressions except for **float** or **double** types and the addresses of external objects, which includes function names, references and pointer variables. While a full discussion of templates far exceeds the scope of this book, the example below illustrates some of these features:

```
template <class T = int> struct square {
      double operator ( ) ( T aX ) { return aX * aX; }
};

int cube( int x ) { return x * x * x; }

template < typename T, int (*aF) (int) > double test( int aI ) {
      cout << (*aF)( 10 ) << ' ' ;
      return aF( aI );
}

main ( ) {
      cout << test<square<double>, cube > ( 6.0 );
}                                          // Output: 1000 216
```

Template classes can simplify program structure and hasten execution. To create a template class, a specialized case should be first coded and verified and only then subsequently generalized by successively introducing template parameters.

14.11 The complex class

Complex numbers are implemented as templates through the **<complex>** header file. Thus a complex-number object with **double** real and imaginary values is defined as

```
complex<double> c;
```

To initialize a **complex** object to a value such as $1 + 2i$, either of the following two statements can be employed:

```
complex<double> c = complex<double>( 1., 2. );
complex<double> c( 1., 2. );
```

The real and imaginary parts of **c** can be accessed through the **real( )** and **imag( )** member functions, e.g. **c.real( )** and **c.imag( )**. Standard arithmetic operators such as $+, -, *, /, +=\ldots$ as well as the stream insertion and extraction operators << and >> are overloaded for **complex** objects. Additional functions in the **complex** class include **arg( )**, **conj( )**, **abs( )**, **polar(r, t)**, which yields re^{it}, where r and t must be floats or doubles, **cos( )**, **cosh( )**, **exp( )**, **log( )**, **log10( )**, **pow( )**, **sqrt( )**, **sinh( )**, **tan( )** and **tanh( )**. Hence

```
#include <complex>
main( ) {
    complex<double> c1 = complex<double>( 1., 2. ), c2( -1., 1. );
    complex<double> c3 = ( c1 + c2 ) / 3.0;
    c1 = exp( M_PI * c3 );
    c2 = polar( 1., M_PI / 2. );
    cout << c1 << '\n' << c3 << '\n' << c2 << endl;
}
```

yields

```
(-1,1.22461e-16)
(0,1)
(6.12303e-17,1)
```

14.12 The standard template library

Modern C++ compilers generally include the STL (standard template) data-structure library. While a detailed discussion of these classes exceeds the scope of this text, several frequently recurring features illustrate the relevant interfaces.

The fundamental data types implemented by the STL library are **vector**, **dequeue**, **list**, **set**, **multiset**, **map**, **multimap**, **stack**, **queue** and **priority_queue**. Each is defined through a corresponding **#include** statement, e.g. if a **stack** appears in a program, its definition must be preceded by the statement **#include <stack>**. However, **queue** and **priority_queue**, **map** and **multimap** and **set** and **multiset** share the include files **<queue>**, **<map>** and **<set>**, respectively.

Although the member functions of each of the above classes differ, certain functions are common to almost all STL classes. These include the destructor, copy constructor and assignment operator, the **empty( )** function that returns **true** if the container is empty, the **max_size( )** function that can be used to set the maximum container size, the **size( )** function that returns the number of elements in the container, **erase( )** and **clear( )**, which erase a given number and all elements from the container, respectively, and the comparison operators **<, >, <=, >=, ==** and **!=** that compare the elements of two similar classes. Another shared concept is the **iterator**, which is an object that can be used to step through the elements of a container (except for **stack**, **queue** and **priority_queue**). An **iterator** object, defined for a **vector** container through the syntax

```
vector <object type>:: iterator it;
```

possesses two member functions, **it.begin( )** and **it.end( )**, that respectively return a pointer to the first member of the container and to a fictitious member one element beyond the end of the container. The iterator class includes an increment operator $(++)$ and a decrement operator $(--)$ that displace the pointer by one container element, the comparison operators $==$ and $!=$ and the assignment operator, =. The iterators of the random-access **vector** and **deque** classes further permit random access in the container through the index operator **it[n]**, the **at** member function **it.at(n)** or the corresponding pointer expression *(**it** + **n**).

We illustrate the above concepts through a simple example that stores and then prints four values {0, 1, 2, 3} in a four-component STL **vector** object (note that the elements of an STL object are initialized to zero when defined):

```
#include <vector>
#include <iterator>

main ( ) {
      // Elements initialized to zero
      vector < int > aV( 3 );
      vector < int > :: iterator pItV;
      // Size expanded to 4 elements; aV[3] = 3.
      aV.push_back( 3 );
      aV[1] = 1;
      aV.at( 2 ) = 2;
      for ( pItV = aV.begin( ); pItV < aV.end( ); pItV++ )
            cout << *pItV << endl;
}
```

The output of this program is

```
0
1
2
3
```

The **push_back(3)** statement adds an additional element **aV[3] = 3** to the end of the vector **aV**. The **at( )** function, unlike the index operator **[ ]**, implements bounds checking so that attempting to address a nonexistent vector element raises a runtime exception. Other important functions are the **sort( )** function of the **vector** and **dequeue** classes, and the **insert( )**, **remove( )** and **resize( )** functions of the **vector** class that respectively insert or remove values at specified locations and resize the vector to a user-specified value. The **resize( )** function can be employed to avoid the automatic resizing of the **push_back( )** operation, which either increases the size of a vector by one for each invocation or doubles its size.

The data types **set**, **multiset** and **priority_queue** sort values after insertion according to a comparator functor, which is a class employed in the same manner as a function. For simple built-in data types, the default comparator automatically sorts the values in increasing order as for the set below (the constructor arguments for **S1** denote the address for the pointer **S1.begin( )** and for **S1.end( )**, respectively, where the last pointer is one element beyond the end of the construct):

```
#include <set>
#include <iterator>

main ( ) {
      double a[3] = {  1.3, 2.5, 0.3 };
```

```
        set < double > S1( a, a + 3 );
        set < double > :: iterator pItS;
        S1.insert( 0 );
        S1.insert( 1.3 );
        for ( pItS = S1.begin( ); pItS != S1.end( ); pItS++ )
            cout << *pItS << endl;
}
```

This yields

```
0
0.3
1.3
2.5
```

For user-defined data types, however, a comparator functor, when required, must be specified as in

```
#include <set>
#include <iterator>

class C {
    public:
    int value1, value2;
};

class myGreater {
    public:
    myGreater( ) {
    }
    bool operator( ) ( const C &C1, const C &C2 ) {
        return( C1.value1 < C2.value1 );
    }
};

main ( ) {
    C C1 = { 1, 2 };
    C C2 = { 3, 4 };
    set <C, myGreater> s;
    set <C, myGreater> :: iterator pItV;
    s.insert( C1 );
    s.insert( C2 );
    for ( pItV = s.begin( ); pItV != s.end( ); pItV++ )
        cout << (*pItV).value1 << ' ';
}
```

which yields 1 3. The principle of operation of the functor is that, since the constructor has no effect in the class **myGreater**, the function **myGreater(C1, C2)** is invoked without first instantiating a **myGreater** object. The member operator function **operator( )** of **myGreater** is then called directly.

Finally, the STL contains mathematical functions that operate on STL data structures as in

```
#include <algorithm>
#include <iterator>
#include <vector>
```

```
// for accumulate
#include <numeric>

int tripleFunction ( int aValue ) { return 3 * aValue; }
int tripleAccumulateFunction( int aPartialSum, int aValue ){
      return aPartialSum + 3 * aValue;
}

main( ) {
ostream_iterator <int> myOut( cout, "loop element \n" );
vector < int > myVector ( 20 );
// places 5 in all 20 positions
fill( myVector.begin( ), myVector.end( ), 5 );
// changes first three values to 3
replace( myVector.begin( ), myVector.begin( ) + 3, 5, 3 );
// three-element zero vector
vector < int > tripleResult( 3 );
// third and fourth elements copied and tripled
transform( myVector.begin( ) + 2, myVector.begin( ) + 4,
      tripleResult.begin( ), tripleFunction );
// new vector sorted
sort( tripleResult.begin( ), tripleResult.end( ) );
copy( tripleResult.begin( ), tripleResult.end( ), myOut );
cout << endl;
// sums elements
cout << accumulate( tripleResult.begin( ), tripleResult.end( ), 0 );
cout << endl;
// triples and sums
cout << accumulate( tripleResult.begin( ), tripleResult.end( ), 0,
      tripleAccumulateFunction );
}
```

which yields

```
0 loop element
9 loop element
15 loop element
24
72
```

In the program, an **ostream_iterator** object **myOut** is first created for **int** values. Iterators of this type are associated with **cout** so that the **copy** function can be employed to print out the elements of the container (here a **vector**) by sending these to the **ostream_iterator**. A 20-element and a 3-element **vector** object, which store **int** values, that are by default initialized to zero are defined. Subsequently, all elements of the 20-element vector, **myVector**, are set to 5 and then 5 in the first three elements is replaced by 3. The third and fourth elements in **myVector** are tripled through the user-defined function **tripleFunction( )** that is passed as a parameter to the STL **transform( )** function, which places the result into the first two positions of the three-element zero vector **tripleResult**. The three elements are sorted and the result printed. Finally, two forms of the **accumulate( )** function process the elements of the **tripleResult** vector. The first simply adds the elements while the second calls the user-supplied global function **tripleAccumulateFunction( )** to evaluate the sum of three times each element.

The **transform( )** function accepts certain predefined functors as arguments, such as in

```
transform( inputVector1.begin( ), inputVector1.end( ),
    inputVector2.begin( ), outputVector.begin( ),
    minus<double>( ) );
```

which subtracts the elements of **inputVector2** from the corresponding elements of **inputVector1** and places the result into **outputVector**. Many other functors can replace **minus<double>( )**, such as **plus<double>( )** or **multiplies<double>( )**; and the output vector can be identical to either of the input vectors. The vector inner product is

```
inner_product( inputVector1.begin( ), inputVector1.end( ),
    inputVector(2).begin( ), 0., plus<double>( ),
    multiplies<double>( ) );
```

14.13 Structures, unions and nested classes

A structure is identical to a class, except that the internal data members are public rather than private by default, and the keyword **class** is replaced by **struct**. Thus

```
class MyClass {
    int m;
    public:
    double n:
}
```

is functionally identical to

```
struct MyClass {
    double n;
    private:
    int m;
}
```

A **struct** often groups public variables of different types into a generalized array.

A **union** is identical to a **struct** except that all data members share the same storage:

```
union U{
    int m;
    int n;
}

main ( ) {
    U U1;
    U1.m = 3;
    U1.n = 0;
    cout << U1.n << endl;              // Output: 0
}
```

A **union** reduces the memory required by a program, but only a single variable in the **union** can be accessed at a given time. Unions can also simplify access to different memory locations in a data type, as illustrated by

```
union U{
      int m;
      char c[4];
};

main( ) {
      U U1;
      U1.m = 65;
      cout << U1.c[0] << endl;                    // Output: A
}
```

which displays A, corresponding to ASCII code 65. If a union contains two variables, such as an integer and a single-character variable that occupy differing amounts of storage, the variables overlap at the least-significant memory bits.

Class (and structure) definitions can be nested as in the example below:

```
struct S {
      int iS;
      struct C {
            int iS;
      };
      print( ) { C C1 = { 2 }; cout << iS << '\t' << C1.iS; }
};
main( ) {
      S S1 = { 1 };
      S1.print( );       // output: 1   2
      cout << S1.C.iS;   // compile error: C not visible outside S
}
```

However, a nested class is accessible *only* to the members and friends of the class.

14.14   Bit-fields and operators

Applications such as interfacing a computer with external devices often require manipulation of single bits or sets of bits. This process can be facilitated by defining a *bit-field* of the form

```
struct myBitField {
      unsigned firstBit: 1;
      unsigned secondBit: 2;
      unsigned thirdBit: 1;
};
```

Then e.g. **myBitField MB;** creates a one-bit variable labeled **MB.firstBit**, a two-bit variable **MB.secondBit** and a second one-bit variable **MB.thirdBit**. These variables are ordered in either ascending or descending memory locations depending on the computer hardware; however, the first bit appearing in the bit-field is placed in the least-significant bit (bit zero) of the memory space reserved for the field.

C++ provides operators that manipulate individual bits within a given variable. These comprise the bitwise logical operators and (**&**), or (|), exclusive or (^), not (~) and the right and left shift operators >> and << that shift the bit pattern of the variable right and left by a given number of bits (specified to the right of the operator), respectively, and introduce 0 in place of bits that are dropped. The **&** and >> operators are illustrated below through two methods for printing out the bit pattern associated with an arbitrary character variable (the union overlaps the single-bit field with the least significant bit of the character variable):

```
struct aBit {
        unsigned bit: 1;
};

union charBit {
        char c;
        aBit aB;
};

main ( ) {
        charBit CB;
        CB.c = 'e';
        for ( int loop = 0; loop < 7; loop ++ ) cout << " ";
        for ( int loop = 0; loop < 8; loop++ ) {
                ( CB.aB.bit ) ? cout << "1" : cout << "0";
                CB.c = CB.c >> 1;
// alternative code
//      ( CB.c & 1 ) ? cout << "1" : cout << "0";
//      CB.c = CB.c >> 1;
        }
}
```

14.15 Program optimization

Advanced constructs such as templates, virtual function calls, pointers and references restrict the range of optimization techniques available to the compiler, which must retain features that, although rarely employed, could result from these language elements. Proper coding can, however, compensate for the resulting loss of efficiency. For example, often a large fraction of the computation time is spent in a few inner loops. These time-critical code segments can be located and often subsequently streamlined through a profiler such as that available by setting the compiler options in Dev-C++. However, many standard procedures for enhancing program operation are automatically performed by modern C++ compilers, as, for example

- Replacing multiplication by addition, floating-point artithmetic by integer arithmetic and doubles by floats (floats are often automatically converted to doubles).
- Insuring that the matrix indices furthest to the right are incremented in the innermost loops.
- Reversing loops so that the loop index runs backward from the largest value of the iterator to the smallest value.
- Unrolling loops so that a loop containing e.g. 20 iterations of one statement is replaced by a loop containing five iterations of four identical statements.
- Employing the **register** keyword to instruct the compiler to place certain variables into internal CPU memory registers.
- Blocking or tiling loops to fit subcalculations into main or cache memory. While substantial performance improvements can result, such techniques require knowledge of the computer hardware. The performance advantage also decreases as the speed of main memory approaches that of cache memory.

Some methods that will often improve performance are

- Replacing division with multiplication and small integer powers by repeated products.
- Insuring that e.g. a product that is repeatedly evaluated in an inner loop but always results in the same value is moved to an outer loop or outside all loops. Similarly, if the same calculation is repeated in an inner loop, its result should be stored in an appropriate array or matrix variable. As an example:

```
for ( int outerLoop = 0; outerLoop < 100; outerLoop++ ) {
    for ( int innerLoop = 0; innerLoop < 100, innerLoop++ )
        A[outerLoop][innerLoop] =
        sin( 2.0 * M_PI * outerLoop / 100.0 ) *
        sin( 2.0 * M_PI * innerLoop / 100.0 );
}
```

can be replaced with

```
double sinConstant = 2.0 * M_PI * 0.01;
double sinArray[100];
for ( int loop = 0; loop < 100; loop++ )
    sinArray[loop] = sin( sinConstant * loop );

for ( int outerLoop = 0; outerLoop < 100; outerLoop++ ) {
    double sA = sinArray[outerLoop];
    for ( int innerLoop = 0; innerLoop < 100, innerLoop++ )
        A[outerLoop][innerLoop] = sA * sinArray[innerLoop];
}
```

- Improving the speed of numerical algorithms. For complex tasks, this often requires a sophisticated program library. However, for well-behaved problems, compact programs such as those found in this text often perform more rapidly. Such code can be tailored to the specific problem, as, for example, if certain input variables are never altered or all elements of an array are identical.

- Use of appropriate data structures. The efficiency of numerical methods that frequently access data can often be improved by replacing arrays with appropriate data structures, such as representing a sparse array by a linked list.
- Selecting the highest correctly functioning optimization flag. Compilers generally possess optimization flags that require the absence of certain programming structures if they are to function as intended. Consequently, a program should be compiled and run without optimization and the results compared with those obtained at higher optimization levels.
- Inlining functions: small functions that are called frequently during execution should be declared **inline** (recall, however, that functions appearing in the body of a class definition are normally automatically inlined).

Chapter 15
Creating a Java development environment

While scientific programs are less frequently written in Java than in C++, in certain contexts, such as internet applications, the enhanced Java feature set significantly shortens development time. Since many high-level constructs in Java reflect an involved and largely hidden underlying structure that often precludes a description in terms of a compact set of underlying principles, the subsequent discussion focuses on compact code samples that illustrate the most significant aspects of the language. Once a basic understanding of Java has been acquired, specialized programming tasks can often be addressed by extending these samples while consulting a full list of the specialized functions available in the language.

While numerous free integrated Java development environments exist, flags can be set inadvertently, leading to anomalous behavior that often proves difficult to correct. Additionally, the details of the Java file structure which constitutes an important feature of the language are often obscured. This text therefore employs a command-line compiler, the Java DISLIN graphics package and a text editor. Downloading, installing and testing these components is described below.

15.1 Basic setup

To being the process of installing Java, download and install the Java JDK (Java Development Kit) from

http://www.oracle.com/technetwork/java/javase/downloads/index.html

the Notepad++ editor from

http://sourceforge.net/projects/notepad-plus/

and finally DISLIN for Java from

http://www.mps.mpg.de/dislin/server.html

The correct DISLIN distribution is named dl_**_jv.zip for 32-bit Windows, where ** represents the version number. *Install DISLIN into a new directory* **X:\dislinjava**, where **X**, which should in most cases be substituted by **C**, represents the drive (local-disk) letter. Otherwise, files in the C++ DISLIN directory,

X:\dislin, created at the beginning of the C++ section of this book will be overwritten.

Next double click on Control Panel and then double click on the System icon. Click on the Advanced System Settings pushbutton or the Advanced tab and then the Environment Variables pushbutton near the bottom of the pop-up window. In the System Variables listbox find the entry marked Path. Click on this entry to highlight it and select Edit. Right click in the text-entry field labeled Variable Value and with the right-arrow on your keyboard advance to the end of the text string. Without introducing a space, add (after again replacing **X** by the letter of the appropriate drive, C or D)

```
;X:\Program Files\Java\jdk1.6.0_17\bin;X:\dislinjava\win;
```

to the end of the string (be sure to place a semicolon between each directory entry and the next, as above). *Replace, however, 1.6.0_17 in the line above with the corresponding number of the JDK version that you have downloaded.* This number can be found by employing e.g. Windows Explorer to navigate to the **X:\Program Files\Java** directory and examining the name of the JDK subdirectory. Finally, if DISLIN for C++ is not installed, repeat this last step with the Variable Name DISLIN and Variable Value field **X:\dislinjava** and depress OK a third time. If DISLIN for C++ has, however, already been installed, find DISLIN in the list of system variables, highlight it, press Edit and replace the C++ install directory, normally **X:\dislin**, with **X:\dislinjava**. Depress OK a final time to exit the Environment Variables menu page. *However, the DISLIN environment variable now does not evaluate to the directory required for C++.* Consequently, either the preceding step should be reversed when programming in C++ or, before compiling DISLIN C++ programs, enter the command (be sure not to include spaces around the equality sign)

```
set DISLIN=X:\dislin
```

from within the Command Prompt window (see below). You can determine whether the system variables have been properly set afterward by entering e.g.

```
echo %DISLIN%
```

Alternatively, retain the C++ system variable setting and enter **set DISLIN=X:\dislinjava** in the command window before compiling Java DIS-LIN programs.

15.2 Command-line operation

Since the Java implementation installed above is run from within a command window, also termed a shell or command-line interpreter, basic command-line operation is reviewed in this section. The normal procedure for opening a command window is to select Start → Programs → Accessories → Command Prompt from the Start button. Next type

```
cd X:
```

where **X** should again be replaced by the letter of the (logical) drive on which the programs will be stored. To navigate into the root directory of **X:**, type

```
cd \
```

To create a subdirectory (folder), **programs**, of the root directory in which to place programs, enter

```
mkdir programs
```

and then type **cd programs** or equivalently **cd .\programs** to enter this subdirectory (. and .. represent the current and parent directory, respectively). More generally, if a subdirectory is contained within a higher-order directory, **directory1**, of the currently active directory, type instead **cd directory1\directory** or equivalently **cd directory1/directory**. The command

```
dir
```

lists the files in this subdirectory (currently none), while

```
del filename
```

deletes the file named **filename**, which must include the three-letter extension when present, such as program.cpp

```
rename oldfilename newfilename
```

renames a file and

```
copy filename newfilename
```

copies a file. Typing

```
rmdir subdirectoryname
```

from the directory above an empty subdirectory with the name **subdirectoryname** removes it. In specifying file names, the asterisk * can be employed to match any sequence of numbers or letters except for the period that separates the file name from the file extension, while a question mark matches any single valid character. Thus **del *** deletes all files without three-letter extensions, **del *.cpp** deletes all files with the extension .cpp, **del h?.*** deletes all files with names beginning in h and that possess a two-letter file name and **del *.*** deletes all files in the directory.

15.3   A first graphical Java program

To start programming in Java, start Notepad++ and enter

```
import de.dislin.*;
class HelloWorldApp {
     public static void main( String[ ] args ) {
          float x[ ] = { 1, 2, 3 }, y[ ] = { 1, 3, 2 };
          Dislin.qplot( x, y, 3 );
          System.out.println( "Hello World" );
     }
}
```

Save this code in the directory created in the preceeding section as the **.java** source file **HelloWorldApp** by selecting File → Save from the menu bar at the top of the editor or depressing the third (save) floppy-disk icon on the button bar and navigating to the myprograms directory. *Be sure to save the file as type Java source file so that it acquires a .java extension* (it will automatically be saved as **HelloWorldApp.java**). *The file name must be properly capitalized.*

Java commands can now be entered directly into the command-line window opened in the previous section. However, Notepad++ provides a more convenient toolbar item through the Execute → Open current dir cmd menu item. After either selecting this item or opening a command-line window, enter **cd X:** followed by **cd myprograms** inside the resulting command window to navigate to the directory containing the program above, and then issue the command **javac HelloWorldApp.java**. If the program was entered correctly, a new file **HelloWorldApp.class** will be created, as can be verified by typing **dir** or **dir h***. If error messages appear, return to Notepad++, correct and save the program and recompile with **javac**. Once the **.class** file has successfully been created, entering **java HelloWorldApp** generates the message "Hello World" together with a graph.

15.4   DISLIN applet

To display a DISLIN graph from within an applet (a .html page), modify the program of the previous section to read

```
import de.dislin.*;
class HelloWorldApp {
     public static void main( String[ ] args ) {
          float x[ ] = { 1, 2, 3 }, y[ ] = { 1, 3, 2 };
          Dislin.metafl( "JAVA" );
          Dislin.qplot( x, y, 3 );
          System.out.println( "Hello World" );
     }
}
```

Again create the file **HelloWorldApp.class** by typing **javac HelloWorldApp.java** and then **java HelloWorldApp**. However, the **Dislin.metafl("JAVA")** statement now instructs Java to create the file **dislin_1.java** (if a file of this name already exists in your directory, the resulting file is labeled **dislin_2.java**, as will be indicated during program execution). This file contains a Java program for the graph displayed in the previous section. To introduce the graph into a .html page, first create the file **dislin_1.class** by entering **javac dislin_1.java**. Then, returning to notepad++, create the following text file:

```
<HTML>
<TITLE> A Dislin Example </TITLE>
<BODY>
Text to appear before the applet
<P><APPLET codebase="." code="dislin_1.class" width=900 height=700>
Text to appear if an error prevents the applet from appearing
</APPLET><P>
Text to appear after the applet
</BODY>
</HTML>
```

Save this as a **.html** file by selecting File → Save as menu item from the menu bar. To view the .html page select Run → Launch in (the desired browser) on the Notepad++ menu bar. If the page is changed, the browser should be closed and this menu item reselected.

15.5 Graphics applet

Java has an extensive set of built-in graphics routines that can be employed to extend the following sample applet:

```
import java.applet.Applet;
import java.awt.*;

public class myApplet extends Applet {
        private static Frame window = new Frame( "Drawing" );
        public void paint( Graphics g ) {
                g.drawString( "A Graph", 400, 400 );
                g.drawLine( 0, 10, 500, 500 );
                g.drawLine( 100, 100, 40, 600 );
                g.setColor( Color.cyan );
                g.fillOval( 300, 300, 80, 275 );
        }
}
```

Saving this code as **myApplet.java**, compiling with **javac myApplet.java** and modifying the .html file of the previous example by replacing **dislin_1.class** with **myApplet.class** generates a drawing containing graphics components that are apparent from the code above. The important feature of this code is that an applet that employs the abstract window toolkit (**java.awt.***) does not employ a **main( )** function. Instead, the **public void paint(Graphics g)** function is called and the enclosed graphics directives executed. Since such constructs, which cannot be predicted from a set of underlying language principles, occur frequently, the reader is advised at least initially to solve problems by modifying preexisting programs wherever possible.

15.6 Packages

Every Java type must be encapsulated within a class or a related entity, namely an interface, enumeration or annotation (a Java version of a template). *Each Java file must contain a single public class with the same name (including capitalization) as the file but without a .java extension.*

In order to avoid name collisions and thus facilitate dynamic loading of classes at runtime Java replaces the C++ **namespace** with a **package** structure such that every class belongs to a package. While a class name is a single, normally capitalized word, a package name normally consists of several words joined with periods, such as **java.awt**. *This name serves as an effective namespace for each element in the package, so that e.g. an object (reference) of a class* **Printer** *in* **java.awt** *can be defined by* **java.awt.Printer P1;**. The current directory functions as an unnamed default package such that any class file is visible to other classes in the directory. A user-defined package labeled **packageName** is generated by inserting **package packageName;** at the beginning of every file in the package. *These files must further be placed in a subdirectory of the active directory with the same name,* **packageName**, *as the package*. If the file is instead included in a subdirectory, **packageName1**, of **packageName**, then **packageName** must be replaced by **packageName.packageName1** throughout. Suppose a class **MyClass** exists inside the file **MyClass.java** that is in turn situated in the package **packageName** that is a subdirectory of the current directory. Then, a *Java program in the current directory can create an object of type* **MyClass**, *either with the statement* **packageName.MyClass1.MC1 = new packageName.MyClass( );** *or by beginning the program with* **import packageName.*** *to import all files in the package followed by* **MyClass MC1 = new MyClass( );**. A single file in this package is imported through **import packageName.MyFile**. If the class **MyClass** appears also in a second imported package, the full qualifier **packageName.MyClass.MC1** is required when referring to **MC1**. If the packages are located in subdirectories of different directories than the current directory, these directories can be automatically incorporated by setting the **CLASSPATH** environment variable. For example, if **CLASSPATH =.;C:\dir1\packageName** package names are referred either to the current directory (.) or to **C:\dir1\packageName**.

The basic Java library comprises 23 packages. The "java.lang" core language classes are implicitly imported; that is, the Java environment functions as if the statement **import java.lang.*;** were present at the beginning of each program. Classes belonging to java.lang include

```
java.lang.Math              // advanced mathematics functions
java.lang.String            // string-handling functions
java.lang.StringBuffer      // string buffering and
                            // manipulation functions
java.lang.Thread            // thread manipulation and
                            // multithread support
java.lang.Exception         // exception-handling routines
java.lang.Error             // error-handling routines
```

as well as the **Byte**, **Double**, **Float**, **Integer**, **Long** and **Short** classes, which contain functions that act on their respective data types. Additional Java packages that appear in this text include

```
java.applet                 // classes for applets
java.awt                    // GUI classes
java.awt.event              // event classes
java.io                     // input/output classes
java.lang.reflect           // reflection API classes
```

Note that *package* or *import statements must appear first in a program, which must contain a single **public** class definition with the same name as the file*. To illustrate,

Program **PackageExample.java** in the directory X:\rootdirectory

```
//with this present can write Multiply.f( 2, 3 )
//import myFunction.*;
//in place of myFunction.Multiply.f( 2, 3) below
public class PackageExample {
    static private class ExponentialConstants {
        static private double iE = 2.7;
    }

    public static void main( String[ ] args ) {
        myFunction.Multiply.f( 2, 3 );
        System.out.println( ExponentialConstants.iE *
                Multiply.MyConstant.iPi );
    }
}
```

and

Program **Multiply.java** in the package directory X:\rootdirectory\myFunction

```
package myFunction;
public class Multiply{
    static public class MyConstant {
        static public double iPi = 3.14;
    }
    static public void f( int a1, int a2 ) {
        System.out.println( MyConstant.iPi * a1 * a2 );
    }
}
```

The access privileges of entities can be default or **private**, **protected**, package or **public**. *A* **public** *entity is visible from any file, while a package or default* *entity, which is defined without an access specifier, is visible only from within its package (even if it is imported into another program). A* **protected** *entity, which is less restrictive than default, is visible from its package as well as from its subclasses (that can be located in other packages) and, finally, a* **private** *entity is visible only from within its enclosing class.* In the above program the internal class **ExponentialConstants** within **PackageExample** is **private** and can be accessed only by member functions of the class itself. However, the internal members of the class **Multiply** are all declared **public** so that they can be accessed by **PackageExample** from outside the **myFunction** package. If, however, the keyword **public** is removed from **public class Multiply**, the elements of **Multiply**, while declared **public**, acquire the package access of their enclosing class and are no longer accessible to **main( )** in **PackageExample**. *The return type of a function must immediately precede its name, thus* **static public int f( )** *is valid but not* **int public f(void)**.

15.7 Static (instance) and class members

As in C++, a **static** variable is shared by all objects of the same class, while class variables differ between objects. *A static method can be invoked without creating any objects of its class type and therefore can only access other static methods and variables directly or non-static methods and variables through an existing object.* Thus, since **f( )** is a **public static** function in **Multiply** above, *it can be called in* **PackageExample** *through the syntax* **Multiply.f( )**, *even though no objects of type* **Multiply** *have been defined. Non-static instance methods, on the other hand, can access both static and non-static member variables.*

In practice this implies that, to access non-static variables from within a static function such as **main( )**, *an object of the class – which can also include the non-static variables – containing the function must be created.* That is, to implement the above program with non-static in place of static variables requires

```
import myFunction.*;
class ExponentialConstants {
      double iE = 2.7;
}

public class PackageExample {
     ExponentialConstants iEC1 = new ExponentialConstants( );
     public static void main( String[ ] args ) {
           PackageExample PE1 = new PackageExample( );
           Multiply M1 = new Multiply( );
           M1.f( 2, 3 );
           System.out.println( PE1.iEC1.iE * M1.iMC1.iPi );
     }
}
```

and in the myFunction subdirectory

A Java development environment

```
package myFunction;
public class Multiply{
     public class MyConstant {
          public double iPi = 3.14;
     }
     public MyConstant iMC1 = new MyConstant( );
     public void f( int a1, int a2 ) {
          Multiply M1 = new Multiply( );
          System.out.println( M1.iMC1.iPi * a1 * a2 );
     }
}
```

Chapter 16
Basic Java programming constructs

While in Java all code must be contained within an encompassing class definition, inside this class the code can closely resemble a procedural program. Accordingly, we first examine these basic features, postponing a detailed discussion of classes and objects until the subsequent chapter.

16.1 Comments

Text to the right of the delimiter // is ignored by the compiler, as is any text enclosed between the starting and terminating delimiters /* and */, as in

```
int m = 10;        // Comment 1
int n = 10;        /* Comment
        2 */
```

Additionally, text between the starting delimiter /** and ending delimiter */ is employed by the javadoc utility included with the Java runtime library to generate HTML documentation. Some relevant tags are

```
/**
@ author            (author of a class)
@ version           (version of a class)
@ see               (link to related topic)
@ return            (method return value)
@ param             (method parameters)
@ exception         (exception thrown by a method)
*/
```

The @ must be located in the first column unless a star (*) is present in this column.

16.2 Primitive types

A variable identifier (name) is composed of any letter, number, underscore or dollar-sign character followed by any combination of letters, digits or underscores. Java is case-sensitive, so the variables **myinteger** and **myInteger** are different. Variables are either primitive (atomic) or non-primitive class types. Primitive types, which are largely patterned on C++ built-in types, are optimized for repeated manipulation such as loop iterations and are accordingly defined and initialized through statements of the form **int loop = 3;**. If an initializer is not supplied, any attempt to utilize the variable will be flagged by the compiler until a meaningful value is assigned.

An **int** always occupies 4 bytes = 32 bits of storage (although primitive data types can be allocated more storage than required to store their value) and therefore represents numbers between $-2^{31} = -2,147,483,648$ and $2^{31} - 1 = 2,147,483,648$. If an **int** overflows or underflows these bounds, only the least-significant bits are retained, which implies that *adding one to the highest positive integer yields the lowest negative integer*. A **long** integer instead utilizes 8 bytes. Similarly, a **byte** corresponds to an 8-bit integer between $-2^7 = -128$ and $2^7 - 1 = 127$, while a **short** contains 16 bits. *A* **boolean** *variable can be set to either of* **true** *or* **false** *and is not automatically converted to an integer*.

Floating-point types in Java encompass a 32-bit **float** that represents a real number between $-3.0E38$ and $3.0E38$ and a 64-bit **double** variable ranging from $-1.8E$-308 to $1.8E308$. A float constant is represented as 10.F0 or 10f, while the corresponding double constant is 10., 10.E0 or 10.e0. A number that underflows these bounds is set to zero; one that overflows the bounds yields **POSITIVE_INFINITY** or **NEGATIVE_INFINITY** and the result of an undefined arithmetic operation such as division by zero is set to **NaN** (not a number).

A **char** represents a 16-bit Unicode international character, which is written as \u**xxxx**, where **x** is a hexadecimal digit (0–9 or A–F). The first 256 characters **0x00** (\u**0000**) to **0xFF** (\u**00FF**) coincide with the 256-bit ASCII character sequence. A character specified by **0yy**, where **y** is an octal digit (0–7) expresses an extended C or C++ character sequence. All such characters must be enclosed in single quotation marks. Thus an ASCII value such as 3 (which is 51 in ASCII or 63 in octal or 33 in hexadecimal) can be written as **0x33** (without quotation marks) **'\063'**, **'\63'** or **'\u0033'** or simply in character representation as **'3'**. As in C or C++, important non-printing Java characters are

newline	\n
horizontal tab	\t
vertical tab	\v
backspace	\b
carriage return	\r
formfeed	\f
question mark	\?
single quote	\'
double quote	\"

If the keyword **final** is included in the definition of a variable, it cannot be altered when visible (throughout its scope). Such variables must be initialized when declared, since they could otherwise not be reassigned to a meaningful value, as in

```
final int m = 36;
m = 29;                    // Error: m cannot be changed
```

16.3 Conversions

Java automatically performs widening conversions from smaller to larger variables of compatible type. Since all numeric types are considered compatable,

```
byte n = 10;
int m = n;
```

initializes **m** to 10. However, *converting a larger to a smaller length requires an explicit cast* such as (the C++ cast syntax **int(m)** is not implemented in Java)

```
double m = 10.5;
int n = (int) m;
```

If the range of **m** is larger than that of **n**, the value placed in **n** is the remainder when **m** is divided by the range of **n**, whereas if **m** is a floating-point type and **n** is an integer, any fractional component of **m** is discarded. Hence **n** above equals 10. Conversions occur automatically within expressions such that, if an operation involves variables of two types, the smaller type is automatically promoted to the larger type. *Narrowing conversions* of e.g. a **double** to an **int** are, however, *precluded even in assignment statements*. For example, if **d** is a **double** and **m** an **int**,

```
m = 3 / 2;          // automatically truncated by removing the
                    // decimal part to m = 1
m = -3 / 2;         // rounded in a similar fashion to m = -1
d = -3 / 2;         // truncated and then converted to d = -1.0
d = -3. / 2;        // yields d = -1.5
m = 3. / 2;         // invalid
```

Consequently, to avoid confusion, smaller variable types should be cast into larger variable types before evaluating such mathematical expressions.

Despite the above rules, an **int** *cannot be employed as an argument to a function that requires a* **double** *parameter unless it is first explicitly cast into a double*, as in **f((double) n)**. Alternatively, the primitive **int** variable can be first transformed into an **Integer** class variable, which possesses a **doubleValue( )** member function that converts its internal data to a double type, i.e. if **m** represents an **int**

```
double d = ( (Integer) m ).doubleValue( );
```

byte and **short** variables are automatically converted to **int** in arithmetic expressions. Hence

```
byte b = 1;
byte c = b * b;
```

yields a compiler error because the **int** value of **b * b** cannot be narrowed to a **byte**. Instead,

```
byte c = (byte) ( b * b );
```

is required.

16.4   Operators

Arithmetic operations in Java include +, -, *, / and, for integer **a** and **b**, the remainder operator, **a % b**, which yields the remainder of **a** divided by **b**, such that −3 % 2 is −1. Every operator possesses a precedence level and an associativity rule. For example, the precedence of * and / exceeds that of + and −, so that, in an expression containing both types of operators, * and / are evaluated first. If the expression instead contains a sequence of operators with the same precedence, the order of operation is determined by associativity. For example, = is right-associative, so that **a = b = c;** means **a = (b = c);**. *In general, operators are left-associative except where this generates logical inconsistencies. Thus,* **5 / 6 * 7** *is evaluated as* **(5 / 6) * 7,** *not as* **5 / (6 * 7),** which causes numerous programming errors.

As in C++, the basic precedence rules are

parenthesis and index operators	( ) [ ] .
unary operators	++ − +
cast	**(typename)**
multiplicative	* /
additive	+ −
string concatenation	+
relational	< > <= >=
equality	== !=
logical operators	&& ||
assignment	= += −= *= /=

Further, **a += b;** sets **a** to **a + b** while **a = a − ++b;** first increments **b** and then subtracts this new value from **a**, while **a = a − b++;** first subtracts the initial value of **b** from **a** and only subsequently increments **b**. It should be remarked that in Java fixed rules exist for the order in which operators are evaluated within a statement, which insures the validity of expressions such as **++m + m++** that are ambiguous in C++.

Logical operators return a **boolean** value, which can, if desired, be converted into the integer 0 or 1, through e.g. the ternary if-then-else operator, **integerValue = (booleanValue) ? 1 : 0;**. Logical operations comprise < >, <=, >=, ! (not), **&&** or **&** (and), || or | (or), == (logical comparison) and ^, which represents exclusive or. The |, **&** and the ||, **&&** operators differ in that for the single-character operator, the expressions on both sides of the operator are evaluated, while for the double-character operators (as in C++) whether the right side of the operator is evaluated depends on the logical value of the left side. Except for primitive types and some specific non-primitive types such as **Integer**, for which ==compares values as in C++, the ==operator returns true only if its two operands refer to the same object. This *does not imply*, however, that the objects contain the same values. Normally, to compare the contents of two objects, the **equals( )** class method of the object must be called. For user-defined classes this requires overriding the **equals( )** method of the **java.lang.Object** class. Finally, the **instanceof** operator returns **true** if its left argument belongs to or is derived from a class that either possesses the same class type as or implements the interface of the right argument.

Operators that act on the individual bits stored at a memory location and can be applied to all integer types e.g. (**long**, **int**, **short**, **char** and **byte**) are the complement ~ that interchanges 0 and 1 bits, << and <<<, left shift with and without sign extension, and |, **&** and ^. Integer values in Java are signed with a 1 in their highest-order bit indicating a negative sign. Shifting these values with the >> operator preserves this highest-order bit and thus the sign of the value, while the >>> instead places a zero in the highest-order bit. Further **short**, **char** and **byte** values are promoted to 32-bit **int** values before these operators are applied.

16.5 Control logic

Program flow can be regulated by logical conditions. A **for** statement possesses an initialization statement followed by a logical (boolean) condition and finally an iterator:

```
int sum = 0, sum2 = 0;
for ( int loop = 0; loop <= 10; ++loop ) {
        sum += loop;
        sum2 += loop * loop;
}
```

In all control expressions, the subsequent braces can be omitted if they enclose a single statement, although braces are frequently erroneously omitted when several statements are under the control of the logical construct. The variable **loop** above is defined by the first of the three statements in the **for** structure, cannot be subsequently redefined within the loop and is destroyed when the body of the loop is exited.

A **for** loop can be replaced by the **while** statement:

```
int loop = 0;
while ( loop < 10 ) {
        sum += loop;
        loop++;
}
```

as well as a **do . . . while** loop,

```
int loop = 0;
do{
        sum += loop;
        loop++;
} while ( loop < 10 );
```

However, statements under the logical control of a **do . . . while** construct are executed at least once, even if the **while** condition is initially false.

An **if** statement can be followed by an **else** statement, as in

```
if ( logical expression ) {
        statements1
}
else {
        statements2
}
```

which is abbreviated by the ternary operator

```
logical expression ? { statements1 } : { statements2 }
```

Control can be passed out of a running loop through the **continue** and **break** statements. These can contain a label, permitting control to be transferred to the end of an enclosing labeled block (which for **continue** must be an enclosing loop):

```
label1: for ( int outerLoop = 0; outerLoop < 20; outerLoop++ ) {
        for ( int loop = 0; loop < 30; loop++ ) {
                if ( loop * loop < 5 ) continue;
                else if ( loop * loop = 125 ) break;
                else if ( loop % 25 ) continue label1:
                // The first continue transfers control here,
                // i continues to increment
        }
// break transfers control to the first statement outside the
// loop in which it appears continue transfers control
// to the end of the block label1
}
```

Conditional branches can also be implemented through a switch statement:

```
switch ( loop ) {
        case 1: { statements1 } break;    // executed if loop == 1
        case 2: { statements2 } break;    // executed if loop == 2
        default: { statements3 }
}
```

The **break** statements pass control to the first statement following the **switch** block; otherwise, if present, the optional **default** statement will always be executed together with the statements following all logical conditions that are satisfied.

16.6 Enumerations

An **enum** type can be assigned only specific alphanumeric values. The syntax of this construct is illustrated by

```
enum enumName {a, b, c};
public class test {
        static public void main(String[ ] args ) {
                enumName myEnum = enumName.b;
                System.out.println( myEnum );
        }
}
```

The only valid assignments to the **enumName** type variable **myEnum** are **enumName.a**, **enumName.b** and **enumName.c**.

Chapter 17
Java classes and objects

A class groups (encapsulates) a set of variables, termed fields in Java, together with the functions, or methods, that manipulate these variables as well as providing access levels such as **private** or **protected** that hide information regarding its components to preclude undesired changes from elsewhere in the program. Class fields generally allow private or protected access and can then be read or changed only through the so-called external interface of the class (and possibly subclasses), consisting of public member functions that access these variables and therefore regulate the permitted operations on the data from outside the class. A class can be viewed as establishing a new data (variable) type that can be employed in a similar fashion to a preexisting Java (reference) type. For example, a function operates on a user-defined object variable argument in precisely the same manner as any Java class argument such as **Integer**.

17.1 Class definition

The following program illustrates the central features of Java class definitions:

```
class MyClass {
    private int iVariable1 = 6;
    public static int iVariable2;
    public int getIVariable1( ) { return iVariable1; }
    public void setIVariable1 ( int aVariable1 )
        { iVariable1 = aVariable1; }
    MyClass ( int aVariable1 ) {
        iVariable1 = aVariable1;
    }
    MyClass( ) { }
}

public class Test {
    static public void main( String args[ ] ) {
        System.out.println( MyClass.iVariable2 );   // Output: 0
        MyClass MC1 = new MyClass( );
        MC1.setIVariable1( 3 );
```

```
        System.out.println( MC1.getIVariable1( ) );   // Output: 4
    }
}
```

Since no specifier is present in the definition of **MyClass**, the class acquires package access; that is, the class and its members are at most visible to files contained in the same package (even including the apparently **public** variable **iVariable2** and functions **getIVariable1** and **setIVariable1**). Additionally, since a **package** statement is absent from the beginning of the program, it executes in an unnamed package containing the **main( )** program. Two types of variables are defined in **MyClass**, namely a class variable **iVariable1** and a **public static** variable **iVariable2**. Variables that are **static** exist in a single copy independently of whether any class objects are ever defined, and possess a lifetime that coincides with that of the running program. Consequently, a **static** variable can be accessed through the class name **MyClass** before the creation of any class objects in the first line of **main( )**. Unlike $C++$, the class variable **iVariable1** can be initialized to 6 through a so-called *synthetic constructor* within the definition statement in the class body. While **iVariable1** is **private** and thus inaccessible outside the **MyClass** class definition, **public** get and set member functions that enable the value of this variable to be read and written to from any external or internal program element within the package are provided (again these functions, while declared **public**, acquire the package access of **MyClass**). Associating the names **getIVariable1** and **setIVariable1** for the get and set member functions of a variable **iVariable1** is highly recommended since these coincide with Java conventions and therefore permit the code to be introspected (analyzed) by certain Java toolboxes.

MyClass contains two constructor functions that provide a mechanism for generating objects of the **MyClass** type. These are the single-argument constructor **MyClass (int aVariable)** and the default constructor **MyClass( )**. *If no constructors are defined, Java supplies a default constructor that initializes all internal variables that are not assigned values in the class definition to zero or NULL. However, if a constructor with any number of non-zero arguments is defined, a default constructor is absent unless also one is explicitly introduced into the program. Therefore, if any non-zero argument constructor is defined in the class body, the statement* **MyClass MC1 = new MyClass( );** *generates an error unless a zero argument constructor is also explicitly provided as above.*

Every class contains an implicit reference variable, **this**, that refers to the enclosing object, e.g. if **iFieldV** is a class member variable, in the statement

```
void setField( int aFieldV ) { this.iFieldV = afieldV; }
```

this.iFieldV = afieldV; is equivalent to **iFieldV = afieldV;**.

Unlike C++, variables and functions within a Java file can be placed in any order, since Java generates an internal list of the elements before compilation. Further, **javac** checks all dependences on external files and will compile or recompile any required external .java files that are either uncompiled or newer than their corresponding .class files.

17.2 Inheritance

A class, **DerivedClass**, defined with the syntax

```
class DerivedClass extends BaseClass { ... }
```

acquires (inherits) all **public** *and* **protected** *internal variables and functions in the parent, base class,* **BaseClass**, *except for elements that are explicitly redefined in the derived class. A derived class also inherits all of the variables and functions of a parent class with default (package) access, assuming that the two classes are located in the same package.* Additional variables and functions are then provided in the derived class. In general, the relationship of a derived class to its parent class represents an "*is-a*" specialization between two physical objects such as "a pencil is a writing instrument". A "*has-a*" relationship, typified by "a pencil has an eraser" is termed containment, and is implemented by one class possessing an internal class member of a different class type.

    A class member whose definition is preceded by the keyword **final** *cannot be overridden in a subclass. If the class definition is preceded by* **final**, *it cannot be subclassed. Final class variables (internal member variables) must therefore be assigned values either in class constructors or, more commonly, through synthetic constructors in the member variable list.*

    The **super** *keyword refers to the immediate parent in a class hierarchy.* For example, the following procedure modifies (overrides) a parent class **print( )** method while still exploiting its functionality:

```
public void print {
      super.print( );
      System.out.println("In subclass");
}
```

In a similar manner, a superclass constructor can be included in a derived class constructor. The constructor below adds an additional class variable **iFieldK** to those of the superclass:

```
public MySubclass( int aFieldJ, int aFieldK ) {
      super( aFieldJ );
      iFieldK = aFieldK;
}
```

17.3 Java references and functions

In the example in the previous section a **MyClass** object is created and initialized through the statement **MyClass MC1 = new MyClass( );**, which can equivalently be written as

```
MyClass MC1;
MC1 = new MyClass( );
```

Memory for an object is allocated during program execution through a request to the operating system performed by the **new** *operator*. The operating system allocates the appropriate amount of memory and passes the address of this memory back to the running program. This address is then stored in the object *"reference" variable* **MC1**. Reference variables, which encompass all non-primitive Java variables, correspond to *invisibly dereferenced non-constant pointers*. That is, although **MC1** in fact stores the address of a memory location, Java automatically returns the value at the memory location rather than the address. Therefore, a Java program can manipulate and inspect the properties of objects and built-in variables but cannot access their stored memory addresses. *However, since* **MC1** *is in fact a non-constant pointer, it can be assigned subsequently to a new memory address* by, for example, **MC1 = new MyClass( );** or **MC1 = MC2;**, where **MC2** is a second **MyClass** object. Since Java reference variables are effectively pointers and store memory addresses, two variables can access the same memory address and will then behave identically, as in

```
MyClass MC1 = new MyClass( 2 );
MyClass MC2 = MC1;
MC2.setVariable1( 4 );
System.out.println( MC1.getVariable1( ) );          // Output: 4
```

Any reference variable that does not point to an object of its specified type is assigned a **null** *value so that attempting to access its internal members leads to an error condition*. To simultaneously avoid memory leaks, *memory that is dynamically allocated once an object has been created through a* **new** *statement is deallocated after the system has determined that no references to the object exist*. This automatic reclaiming of memory – which occurs at periodic time intervals – is termed garbage collection and can be forced through the statements

```
Runtime R = Runtime.getRuntime( );
R.gc( );
```

The amount of free memory can similarly be obtained by calling **R.freeMemory( )**. Additional actions, such as closing open files, that should occur when an object of a given class is no longer employed can be included in a **finalize** method

```
void finalize ( ) {
  ...
}
```

within the class definition.

When a function is called in Java, new memory space is reserved for the variables defined in the argument list and in the function body. That is, the first statement that is implicitly executed when a function such as **f(typeName aT)** is called through a statement **f(pT)** is to execute the definition and initialization **typeName aT = pT;** followed by the statements contained in the function body. Memory is allocated for all variables defined in the function block and this local memory is deallocated when the block is exited, as with any other block in the program. Hence *a Java primitive is passed by value to a function as illustrated below*:

```
public static void f ( int aI ) { aI = 4; }
...
int k = 3;
f( k );
System.out.println( k );                          // Output: 3
```

Accordingly, when a Java object "reference" is passed, the address of the object is copied. Thus interchanging two objects or assigning a new object to a reference variable inside a function leaves the parameters in the calling program unchanged:

```
public static void f ( Integer aI ) { aI = new Integer (4); }
...
Integer k = new Integer( 3 );
f( k );
System.out.println( k );                          // Output: 3
```

For the same reason, if a value in the function reference argument is altered, it changes in the calling program:

```
class Coord{ int fieldX; }
class Test {
static void newValue( Coord aC ) { aC.fieldX = 20;}

public static void main(String[ ] args) {
    Coord C1 = new Coord( );
    C1.fieldX = 0;
    newValue( C1 );
    System.out.println( C1.fieldX );              // Output: 20
    }
}
```

Note again that the function **newValue** must be declared **static** since the **static main( )** method cannot address a non-static function without first generating an object of type **Test**. However, the static function **newValue** can change the internal variables of **aC** since it is passed a reference to a preexisting object through the parameter list.

The standard mathematics functions with minor modifications are accessible by prefixing the name of the function with **Math**, *as for example* **Math.exp(doubleValue)** *and* **Math.pow(a, b)**, *which is* a^b. *Similarly,* **M_PI** *in Dev-C++ must be replaced by* **Math.Pi**, *while* **M_E** *(the value of e) is replaced by* **Math.E**. The prefix **Math.** can be made superfluous by placing the line

```
import static java.lang.Math.*;
```

at the beginning of the program.

17.4 Exceptions

Java program faults that leave the program in a state that can be further influenced by additional user-provided routines are termed exceptions, while errors (e.g. **VirtualMachineError**) typically cannot be handled by the program. *Java exceptions subclass the* **Exception** *type, through additional inheritance levels of predefined exception types* such as **ArithmeticException**, **SecurityException**, **NoSuchMethodException**, **IOException**, and **NoSuchFieldException**. Any such types can be further subclassed by writing e.g.

```
public class myException extends ArithmeticException { ...
```

If an exception is thrown in a routine either by routines within the Java virtual machine (the Java runtime) or by the program through the syntax **throws ExceptionName**, *the routine either catches it or passes it to the calling program. The latter is accomplished by also including* **throws ExceptionName** *in the signature of the routine.* The calling routine then must similarly either catch the exception or pass it further to the next higher calling level. Multiple catch clauses in a function can perform different operations for each exception type. That is, exceptions can be thrown and handled within a function through a procedure such as

```
try { if ( i == 0 ) throw new Exception( "i = 0 error" );
      if ( i == -1 ) throw new Exception( "i = -1 error" );
}
catch ( Exception e ) { System.out.println( e ); }
finally { i = 2; }              // executed even if no exception
```

where the **try** block can contain any portion of a program including various function calls. If the exception is thrown, the **catch** and **finally** blocks are executed and the program continues with the first statement following the **finally** block. If **continue** or **break** clauses are encountered in control logic constructs, the **finally** clauses are implicitly activated; thus if the following lines are encountered within a function

```
import java.io.*;
try { switch ( k ){
      case 0: throw new IOException( );
      case 1: throw new Exception( );
      case 2: break; }
      System.out.println( "End" );
}
catch ( IOException e ) { System.out.println( "IO" ); }
finally { System.out.println( "F" ); }
```

then for $k = 0$ the code handles the exception so that output is IO and F, for $k = 1$, F is printed and the exception is passed to the caller through a **throws Exception** clause in the enclosing function signature, while for $k = 2$ only F is printed.

17.5 Basic Java reference types

Numerical types. Every primitive data type can be converted into its corresponding reference (class) type by writing e.g. **Byte(b)**, **Double(d)**, **Float (f)**, . . . , where **b**, **d** and **f** are primitive variables of byte, double and float type, respectively. The class types then possess methods such as **doubleValue( )** and **intValue( )**, methods that return the indicated primitive type. Primitives can also be converted to class types through statements such as **int n = 5; Integer M = new Integer(n);** or simply **Integer M = new Integer(0); M = n;** or even **Integer M; M = n;**.

Strings. A **String** object can be created by enclosing a sequence of characters in double quotation marks, enabling constructs such as **"hel\tlo".length( )**, which evaluates to 6, which is the length of the string "hello" together with a single tab character (no termination character is present) and **String S = "hello";**. To input a string over several lines, each line should be preceded by a + sign. *A + sign appearing between a numeric type and a string also converts the numeric type to a string.* Additional **String** functions include **String1.equals(String2)** and **String1 == String2**, both of which return true if the two strings are identical (store the same text), **String1.compareTo(String2)**, which returns zero if the two strings are identical, and **charAt(int characterPosition)**. In addition, **substring(int startIndex, int endIndex)**, **concat(String String2)** and **replace(char originalCharacter, char newCharacter)** return new strings that respectively contain the substring of the original string positioned between **startIndex** and **endIndex**, concatenate the original string with **String2** and replace all instances of **originalCharacter** in the original string with **newCharacter**.

To modify a **String** object directly, as opposed to calling a function that returns a new **String** object, the **String** object should first be placed into a **StringBuffer** through

```
StringBuffer SB1 = new StringBuffer( String1 );
```

which is printed with **System.out.println(SB1);**. The buffer contents are changed through functions such as **insert(int index, String String2)**, which inserts **String2** into the original string at the position of **index**. A string can further be parsed, i.e. divided into a set of tokens (words), by first placing it into a **StringTokenizer** class

```
StringTokenizer ST1 = new StringTokenizer(
    String String1, String Delimiters );
```

where the optional argument **Delimiters** contains the characters that indicate the start or end of each token. Default delimiters include the whitespace characters space, tab, newline and carriage return. Subsequently, each token in the string is obtained with **ST1.nextToken( )**. Strings are converted into the various numerical types by e.g. **Double.parseDouble(S)**, where **S** is a string (a **NumberFormatException** is thrown if the string cannot be interpreted as a numeric value).

Finally, all classes possess **toString( )** functions. However, the default **toString( )** function, if not overridden by a class method, prints out the class name followed by a (generally unique) hexadecimal code (hash code) related to the memory address of the object. To print out meaningful information about an object, each user-defined class should contain a **public** function **String toString( )** that instead prints a formatted description of all internal class variables, calling in this process the **toString( )** function of any internal class member variables of non-primitive, object type. The **toString( )** function is called automatically if the object appears as an argument of, for example, **System.out.println( )**.

Arrays. An array comprises an index-accessible sequence of variables of a similar type that in Java can be both declared and initialized through the allocation of storage space by either of the two equivalent statements

```
arrayType arrayName[ ] = new arrayType[arraySize];
arrayType[ ] arrayName = new arrayType[arraySize];
```

Array elements are automatically initialized to zero or to NULL according to whether **arrayType** corresponds to a primitive or class type, respectively. Arrays are initialized to non-trivial values when defined through the following syntax:

```
int a[ ] = { 0, 1, 2, 3 };
```

To alter the contents of the array after it has been defined, however, requires iteration, as in

```
for ( int loop = 0; loop < 4; loop++ ) a[loop] = 2 * loop;
```

An array is generated without specifying an array name by e.g.

```
int k = intFunction( new int[ ] { 0, 1, 2 } );
```

If arrays are equated, both of them refer to the same memory. Changing an array element through one array name then changes the corresponding element referred to by the second array name,

```
Integer Array1[ ], Array2[ ] = { 1, 2, 3 };
Array1 = Array2;
Array1[0] = 4;
System.out.println( Array2[0] );          // Output: 4
```

Although the length of an array, *which is obtained through* **Array1.length** *rather than* **Array1.length( )**, cannot be changed, objects of the **Vector** class are automatically resized to accommodate element addition or deletion through methods such as **Vector1.addElement(element);** and **Vector1.removeElement(element);** which add or remove an element from the end of the object **Vector1**. A **Vector** object *cannot contain primitives* and is created and manipulated through e.g. the syntax

```
import java.util.Vector;
Vector <Double> Vector1 = new Vector<Double>;
Vector1.addElement ( new Double( 3.5 ) );
System.out.println( Vector1.elementAt( 0 ) );    // Output: 3.5
System.out.println( Vector1.size( ) );           // Output: 1
```

A two-dimensional array possesses three equivalent definitions:

```
arrayType matrix[ ][ ] = new arrayType[arraySizeR][arraySizeC];
arrayType[ ] matrix[ ] = new arrayType[arraySizeR][arraySizeC];
arrayType[ ][ ] matrix = new arrayType[arraySizeR][arraySizeC];
```

and can be initialized when created with the notation

```
int matrix[ ][ ]={ { 1, 2 }, { 2,3 }, ... };
```

Such an array can be passed to a function with e.g. a signature **returnValue myFunction(arrayType aMatrix[ ][ ]);**. A non-square matrix is generated as follows:

```
double [ ][ ] NonSquareMatrix = new double [2][ ];
NonSquareMatrix[0] = new double[2];
NonSquareMatrix[1] = new double[3];
```

17.6   Input and ouput

To enter information into a Java program from the command line when the program is started, further arguments can be supplied to the **java** command after the program name. These are captured as **String** objects by the **args[ ]** argument of **main( )**. Therefore, issuing **java myProg 3.5** enters 3.5 into **args[0]**, which is retrieved as follows:

```
public static void main( String args[ ] ) {
      Double x = Double.valueOf( args[0] );
...
}
```

To enter primitive types from the terminal (or a **String** or any **InputStream**) we use the syntax

```
import java.util.Scanner;                // at beginning of program
Scanner myIn = new Scanner( System.in );
String SpringType = myIn.nextLine( );
double force = myIn.nextDouble( );
myIn.close( );
```

which reads a **String** value followed by a **Double** value.

Normally output is displayed on the standard output device (the monitor) through

```
System.out.println( "Hello World" );
```

which sends the string "Hello World" to the terminal together with a carriage return. Replacing **println** with **print** eliminates the carriage return. Integer numbers are formatted by

```
java.text.NumberFormat NF1 = java.text.NumberFormat.getInstance( );
```

after which

```
System.out.println( NF1.format( 100000 ) );
```

yields the output 100,000. Calling the methods of **NF1** yields alternative output formats. Floating-point number formatting is specified by

```
java.text.DecimalFormat DF1 =
      new java.text.DecimalFormat( "00.####E0" );
double d = -123.45;
System.out.println( DF1.format( d ) );  // Output: -12.345E1
```

which prints three digits after the decimal point, followed by an E and the mantissa. If a 0 appears in the argument of **DecimalFormat**, a zero is inserted if a number is absent in the indicated position, while a # sign instead places a blank character is this position.

17.7   File I/O

Reading and/or writing to a file is most simply accomplished by starting a program with the command

```
java MyProgram < MyInputFile > MyOutputFile
```

MyProgram accepts input from **MyInputFile** in the same manner as from the keyboard and sends output to **MyOutputFile**.

Chapter 18
Advanced Java features

To conclude the discussion of the Java language, several advanced techniques of relevance to scientific programming are surveyed. These include dynamic method dispatch, abstract classes and interfaces, multithreading and reflection.

18.1 Dynamic method dispatch

A derived class object implements an "is-a" relationship and therefore automatically constitutes an instance of its base class. *Thus a derived class object may be employed in any context in which a base class object is required or expected.* However, *if a derived class object that overrides functions of its base class is employed in place of a base class object, the overridden subclass definitions are resolved and applied at runtime* – hence class member functions correspond to C++ virtual functions and Java "reference" variables act in the same manner as C++ pointers. To illustrate,

```
class C {
      public void print( ) {
            System.io.println ( "C" );
      }
}
class D extends C {
      public void print( ) {
            System.io.println ( "D" );
      }
}

public class DispatchExample{
      static public void main( String [ ] args ) {
            D D1 = new D( );
            C C1 = new C( );
            C CArray [ ] = { C1, D1 }
            CArray[0].print( );                    // Output: D
            CArray[1].print( );                    // Output: C
      }
}
```

All objects in Java are subclasses of **java.lang.Object**, which insures the presence of certain common methods such as **toString( )** and **equals( )**.

179

Accordingly, containers that accept this type act as templates for storing any variety of objects, but *they must subsequently be downcasted from* **Object** *to their actual type through an explicit cast in order to access their specialized class properties.*

18.2   Abstract classes

An **abstract** class contains one or more **abstract** functions that lack definitions, as in

```
abstract class C {
      abstract void print( );
}
class D extends C {
      void print( ) {
            System.out.println ( "D" );
      }
}

public class DispatchExample{
      static public void main( String [ ] args ) {
            C C1 = new D( );
            C1.print( );                          // Output: D
      }
}
```

A subclass that does not supply all these definitions must also be marked **abstract**. *An* **abstract** *object cannot be instantiated since all its methods are not fully defined.* However, dynamic method dispatch allows a **C** reference variable to be assigned an instance of **D** at runtime. For example, a call to the **print( )** function through this reference then calls the subclass method. The abstract function declaration thus serves as a template, that is, specifies a generic form, for the structure of the derived classes.

18.3   Interfaces

Interfaces provide an alternative to abstract classes. *An* **interface** *specifies methods that must be implemented in all classes (often functionally unrelated) that are derived from (conform to) the interface.* An interface and its implementing classes generally require default or public access. *Initialized variables in an interface are implicitly* **public**, **static** *and* **final** *since they cannot be changed by these classes, while all interface methods are implicitly* **public** *and must be declared* **public** *in implementing classes*, for example,

```
interface PrintInterface {
      void print( );
      double pi = 3.14;
}
```

```
class D implements PrintInterface {
      public void print( ) {                  // Note: must be public!
            System.out.println ( "D" + pi );
      }
}

public class DispatchExample {
      static public void main( String [ ] args ) {
            PrintInterface PI1 = new D( );
            PI1.print( );                      // Output: D 3.14
      }
}
```

Since interfaces are distinct from classes, two classes that are unrelated in their class hierarchy can implement the same interface. This feature can, for example, be employed to import long lists of constants into any implementing class. All member functions that conform to an interface in a class, e.g. the **print( )** function above, must possess the same signature as the function in the interface. Interfaces can extend other interfaces, as in

```
PrintAndPlotInterface extends PrintInterface {
      void plot( );
}
```

Finally, a class can implement an interface without providing an implementation of every method specified in the interface. In this case, the class and any subclasses that also do not implement all interface methods must be declared as abstract.

An interface provides a mechanism for passing a function name at runtime as an argument to a second function as illustrated below:

```
interface IArgument {
      double fun( double aX );
}
class Square implements IArgument {
      public double fun( double aX ) { return aX * aX; }
}
class Cube implements IArgument {
      public double fun( double aX ) { return aX * aX * aX; }
}

public class myInterface {
      static double evaluate( IArgument aI, double aX ) {
            return aI.fun( aX );
      }
      public static void main( String[ ] argv ) throws Exception {
            int choose = System.in.read( );
            IArgument I1;
            if ( (char) choose == 's' )
                  I1 = new Square( );
            else
                  I1 = new Cube( );
```

```
                        System.out.println( evaluate( I1, 3.0 ) );
      }
}
```

The interface argument of **evaluate( )** "calls back" the implementing class.

18.4   Java event handling

In the Java event model, which is enabled by **include java.awt.event.*;**, events are produced by an object, called the event source, which passes the resulting event object to an event listener by calling a function in its class. The event object is a subclass of **EventObject** and its fields and methods parameterize all relevant information about the particular event. For a source to call a listener, the listener object must conform to a required interface. Each event source maintains a list of registered listeners by employing the **addEventNameListener( )** and **removeEventNameListener( )** functions. To process, for example, mouse events, it is necessary only to override just the method(s) to which the listener should respond as follows:

```
import java.awt.event.*;
public class MyGraph extends Applet implements MouseListener {
      MyGraph( ) { addMouseListener( this ); }
      public void mouseClicked( MouseEvent event ) {
            MyPoint = event.getPoint( );
            repaint( ); }
      public void mouseEntered( MouseEvent event ) { } ...
```

Following this method, the bodies, which may be empty, for *all* of the functions in the **MouseListener** interface must be provided to specify the action, if any, that occurs when a mouse is clicked, enters or leaves an active window and so on.

As an alternative, one can employ an anonymous inner adapter class that implicitly provides empty bodies for the functions in the interface so that only the functions that have non-empty bodies have to be provided:

```
addMouseListener( new MouseAdapter( ) {
      public void mouseClicked( MouseEvent event ) {
            myPoint = event.getPoint( );
            repaint( ); }
      public void mouseExited( MouseEvent event ) {
            System.exit( 0 ); }
      }
);
```

The words **implements MouseListener** are then omitted from the signature of **main( )**. As an example of the first procedure, the following graphics program generates a canvas of size 600 by 800 pixels within a window frame labeled "Drawing". Text is placed on the canvas and the program then draws a cyan oval and a line from a point 100 pixels down and to the right of the upper left-hand corner of the window to the point at which the mouse is clicked. The program terminates when the mouse leaves the canvas area:

```java
import java.awt.*;
import java.awt.event.*;

public class GraphExample extends Canvas implements MouseListener{
      public int iX = 0;
      public int iY = 0;

      private static Frame window = new Frame( "Drawing" );

      public static void main ( String[ ] args ) {
            GraphExample g = new GraphExample( );
            g.setSize( 600, 800 );
            window.add( g );
            window.pack( );
            window.setVisible( true );
      }

      GraphExample( ) { addMouseListener( this ); }

      public void paint( Graphics g ) {
            g.drawString( "A Graph", 400, 400 );
            g.setColor( Color.cyan );
            g.fillOval( 300, 300, 80, 275 );
            g.drawLine( 100, 100, iX, iY );
      }

      public void mouseClicked( MouseEvent event ) {
            iX = event.getX( );
            iY = event.getY( );
            repaint( );
      }
      public void mouseExited( MouseEvent event ) {
            System.exit( 0 );
      }

      public void mouseEntered( MouseEvent event ) { }
      public void mouseReleased( MouseEvent event ) { }
      public void mousePressed( MouseEvent event ) { }
}
```

The above code can be turned into an applet that instead applies the **Mouse-Adapter** interface with the code below. This program also contains a push button that generates an audible sound when pressed, through the code written into the **Beep** class below that is registered with an **ActionListener**:

```java
import java.awt.event.*;
import java.applet.Applet;
import java.awt.*;

class Beep implements ActionListener {
      public void actionPerformed ( ActionEvent Event ) {
            Component C1 = (Component)Event.getSource( );
            C1.getToolkit( ).beep( );
            GraphExample2 g = new GraphExample2( );
      }
}
```

```
public class GraphExample2 extends Applet {
      Point myPoint;
      public void init( ) {
            Button MyButton = new Button( "Beep" );
            add( MyButton );
            MyButton.addActionListener( new Beep( ) );
      }
      public static Frame window = new Frame( "Drawing" );
      public void paint( Graphics g ) {
            addMouseListener( new MouseAdapter( ) {
                  public void mouseClicked( MouseEvent event ) {
                        myPoint = event.getPoint( );
                        repaint( );
                  }
                  public void mouseExited( MouseEvent event ) {
                        System.exit(0);
                  }
            } );
            g.drawString( "A Graph", 400, 400 );
            g.setColor( Color.cyan );
            g.fillOval( 300, 400, 80, 275 );
            g.drawLine( 100, 100, 300, 300 );
            g.drawLine( 100, 100, (int) myPoint.getX( ),
                  (int) myPoint.getY( ) );
      }
      public void stop( ) {
            System.exit( 0 );
      }
}
```

The **init( )** and **stop( )** methods are called when an applet is created and exited, respectively.

18.5 Multithreading

Every Java program starts inside the main thread, which can be accessed through

```
Thread T1 = Thread.currentThread( );
```

Subsequently, writing **T1.sleep(50)**, which can throw an **Interrupted Exception**, pauses the program for 50 ms. Since, however, the **sleep( )** function is static and acts on the thread it is called from by default, **Thread.sleep(50);** is equivalent to both statements above.

Additional threads can be created by extending **Thread** or by implementing the interface. In the first procedure, the **Thread** class is extended and its **run( )** method overridden. A thread is then activated through its **start( )** method. If instead the **Runnable** interface is extended, a single additional function, **public void run( )**, must be defined. This initiates a new thread by instantiating a **Thread** object and calling its **start( )** method, which executes a callback to **run( )**. To illustrate the first procedure, for each of two threads to print a value and then yield so that the other thread can resume,

```
class Thread1 extends Thread {
      public void run( ) {
              for (int loop = 0; loop < 10; loop++) {
                      System.out.println( loop );
                      yield( );
              }
      }
}
class Thread2 extends Thread {
      public void run( ) {
              for (int loop = 0; loop < 10; loop++) {
                      System.out.println( loop );
                      yield( );
              }
      }
}
class TestThread {
      public static void main( String args[ ] ) {
              new Thread1( ).start( );
              new Thread2( ).start( );
      }
}                             // Output: 0 0 1 1 2 2 ...
```

To synchronize the threads so that each thread waits until the second thread has
completed, they must share access to an object that locks the threads:

```
class MyStaticObject {
      static Object O = new Object( );
}
class Thread1 extends Thread {
      public void run( ) {
              synchronized ( MyStaticObject.O ) {
                      for (int loop = 0; loop < 10; loop++) {
                              System.out.println( loop );
                              yield( );
                      }
              }
      }
}
class Thread2 extends Thread {
      public void run( ) {
              synchronized ( MyStaticObject.O ) {
                      for (int loop = 0; loop < 10; loop++) {
                              System.out.println( loop );
                              yield( );
                      }
              }
      }
}
class TestThread {
      public static void main( String args[ ] ) {
              new Thread1( ).start( );
              new Thread2( ).start( );
      }
}                      // Output: 0 1 2 3 ... 0 1 2 3 ... 10
```

18.6   Serialization

An object can be saved and reconstructed through the **serializable** interface. Object serialization refers to the persistent (permanent) storage of an object, normally by writing the object contents to a serialized file. All fields in the object, including fields inherited from its superclasses, are saved in this manner, except for those explicitly marked as transient. If a field refers to another object, such as an object that contains a second object as a class member (a field), the object that is referenced is also serialized. To serialize an instance of a class **MyClass** below, a file output stream is defined to provide a file into which the object will be written. An **ObjectOutputStream** is defined and is passed the file output stream as a parameter. The object is then written by the **writeObject** serialization method of **ObjectOutputStream**:

```
import java.io.*;
class MyClass implements Serializable {
      private double[ ] myArray = { 0, 1, 2 };
      public double getArrayElement( int aI )
              { return 2 * myArray[aI]; }
}

public class SerializableExample {
      public static void main( String args[ ] ) throws IOException {
            MyClass MC1 = new MyClass( );
            ObjectOutputStream OOS1 = new ObjectOutputStream(
                new FileOutputStream( "myfile.ser" ) );
            OOS1.writeObject( MC1 );
            OOS1.close( );
      }
}
```

The object can later be similarly retrieved through

```
import java.io.*;
public class ReadSerializableExample {
      public static void main( String args[ ] ) throws IOException,
              ClassNotFoundException {
            ObjectInputStream OI1 = new ObjectInputStream(
                new FileInputStream( "myfile.ser" ) );
            MyClass MyObject = (MyClass) OI1.readObject( );
            OI1.close( );
            System.out.println( MyObject.getArrayElement( 0 ) );
      }          // Output: 2
}
```

18.7   Generic types

Analogously to C++ templates, certain Java constructs can possess a formal type (meta)parameter that must be specified when creating an object or calling a function. However, arrays of generic types cannot be created. A class that stores and can assign a value to any object as an internal class member is

```
class MyObject<T> {
      T iT;
      void setT( T aT ) { iT = aT; }
}

class Generic {

      <C> void ClassType ( C aC ) {

            System.out.println( aC.getClass( ).getName( ) );
      }
      public static void main( String args[ ] ) {
            MyObject<Double> MO1 = new MyObject<Double>( );
            MO1.setT(20.0);
            System.out.println( MO1.iT );    // Output: 20
            ClassType( new Double( 1.0 ) );
      }       // Output: java.lang.Double
}
```

As illustrated above, functions can also be defined with type parameters. If such a function is called without specifying the type parameter, the type is generally inferred; that is, calling **ClassType(D1)**, where **D1** is a **Double** object above, prints out the name of the class to which **D1** belongs, namely **java.lang.Double**.

Chapter 19
Introductory numerical analysis

The remainder of this text surveys fundamental programs and techniques in numerical analysis and scientific programming. In this introductory chapter, we clarify error-analysis strategies in the context of a particularly transparent example, namely the derivative operator.

19.1 The derivative operator

Although the derivative is often programmed as a function, it, in reality, constitutes an operator. While a function transforms an argument value, x, to an output value $f(x)$, *an operator, O maps a function argument, $f(x)$, into an output function $Of(x)$.* For the derivative, $Of(x)$ corresponds to the unique function $D_x(f(x)) = df/dx$ described by the slope of the tangent to $f(x)$ at each point x. However, the slope cannot be determined numerically from the information at a single point on a curve, since an infinite family of lines passes through each point. If, however, the slope of the tangent is computed from the differences of the function values at two or more closely spaced points, an infinity of representations for the slope of the tangent function will emerge that differ in accuracy. That is, *while the continuous derivative operator is uniquely defined, it is approximated by an infinite family of discrete operators.* A straightforward approximation is presented in calculus textbooks, in which the derivative is defined as the limit

$$\frac{df}{dx} = \lim_{\Delta x \to 0} \left(\frac{f(x + \Delta x) - f(x)}{\Delta x} \right) \equiv \lim_{\Delta x \to 0} D^+_{\Delta x}(f) \qquad (19.1)$$

In numerical analysis, $D^+_{\Delta x}(f)$ is termed the *discrete forward finite-difference operator*, while Δx, which remains finite in numerical computations, is labeled the step size or point spacing. The continuous and discrete operators in Eq. (19.1) differ by an *error term* that normally varies as a polynomial function of Δx. *If the smallest power of Δx appearing in this polynomial is N, the error decreases as $(\Delta x)^N$ when $\Delta x \to 0$ and the algorithmic accuracy is said to be $O(\Delta x)^N$. To* determine N for the derivative approximation of Eq. (19.1), consider the Taylor-series expansion of a continuous and infinitely differentiable function f about the point x:

$$f(x + \Delta x) = f(x) + \Delta x \frac{df(x)}{dx} + \frac{(\Delta x)^2}{2!} \frac{d^2 f(x)}{dx^2} + \cdots \qquad (19.2)$$

Inserting Eq. (19.2) into Eq. (19.1) yields

$$\frac{f(x + \Delta x) - f(x)}{\Delta x} = \frac{df(x)}{dx} + \frac{\Delta x}{2} \frac{d^2 f(x)}{dx} + \cdots \qquad (19.3)$$

which implies, if $d^2 f(x)/dx \neq 0$,

$$\frac{df(x)}{dx} = D^+_{\Delta x}(f) + O(\Delta x) \qquad (19.4)$$

Since the forward difference expression $D^+_{\Delta x}(f)$ is an operator acting on the function f, code that represents this operator should possess a *function* with a prototype such as **double aF(double);** as one argument. Recall from Section 6.16 that a function name is an alias (alternative name) for and thus evaluates to the compiler-assigned memory location of the first instruction of a binary representation of the instructions comprising the function (the function record). This address can be stored in a pointer to a function of a matching type at compile time or runtime. Since the pointer evaluates to the same starting memory address as the function name, it can subsequently be passed to a derivative function as a parameter:

```
double cube( double aD ) {
      return pow( aD, 3 );
}
double linear( double aD ) {
      return pow( aD, 1 );
}
double derivOperator( double aF( double ),
            double aXValue, double aDeltaX ) {
      return ( aF( aXValue + aDeltaX ) - aF( aXValue ) ) / aDeltaX;
}

main ( ) {
      double deltaX = 1.e-1;
      double xValue = 1.0;
      int choice;
      double (*myFunction) (double);          // Stores memory
                                              // address of function
      cout << "Choose a function 1 - cube, 2 - square " << endl;
      cin >> choice;
      switch ( choice ) {
            case 1: myFunction = cube; break;     // Output: 3.31
            case 2: myFunction = linear; break;   // Output: 1
            default: cout << "Incorrect Input - program exiting";
      }
      cout << derivOperator( myFunction, xValue, deltaX );
}
```

19.2 Error dependence

While the above program displays $[(1 + 0.01)^3 - 1]/0.01 = 3.31$, for the derivative of x^3 at $x = 1$ for $\Delta x = 0.1$ in accordance with an $O(\Delta x) \approx 0.1$ error dependence, its result is exact for $f(x) = x$, since the slope of a straight line is determined by any two points on the line. This demonstrates that the numerical accuracy of a procedure cannot be reliably estimated by analyzing specialized cases. Further, although the numerical error can be reduced by decreasing Δx, for $\Delta x \ll x$, truncation errors in $f(x + \Delta x) - f(x)$ become dominant as discussed in Section 6.30. While a fully reliable routine must therefore optimize Δx, for a wide range of Δx values, the centered difference procedure discussed below proves sufficiently accurate.

For a given numerical procedure, the step size required to achieve a desired accuracy can be obtained through *adaptive* methods. Assume that the difference between the result of an $O(\Delta x)^n$ accurate calculation with a step length Δx and the identical calculation performed with two steps of length $\Delta x/2$ is denoted by ε, while the desired error is E. Then, to reduce the magnitude of the error term by the ratio of the desired to the observed error, the appropriate time step, Δx_{new}, if $\varepsilon > E$ is approximately

$$\Delta x_{\text{new}} = \Delta x \left| \frac{E}{\varepsilon} \right|^{1/n} \tag{19.5}$$

19.3 Graphical error analysis

In most numerical programs, an analytic expression for the error does not exist and the dependence of the error on the input parameters must instead be determined empirically. To illustrate, the following analyzes graphically the variation of the error with step size Δx for the forward finite-difference operator (a description of the plot routines can be obtained by navigating to the DISGNU or DISBCC icon on the Programs submenu of the Start button and selecting DISHLP or DISMAN or by typing DISHLP at the command line):

```
main ( ) {
      double deltaX = 1.0e-1;
      double xValue = 1.0;
      float x[10], y[10];            // array definitions
      for ( int loop = 0; loop < 10; loop++ ) {
            y[loop] = derivOperator( cube, xValue, deltaX );
            x[loop] = deltaX;
            deltaX /= 2;
      }
      metafl( "XWIN" );              // write to terminal
      disini( );                     // start plotting program
```

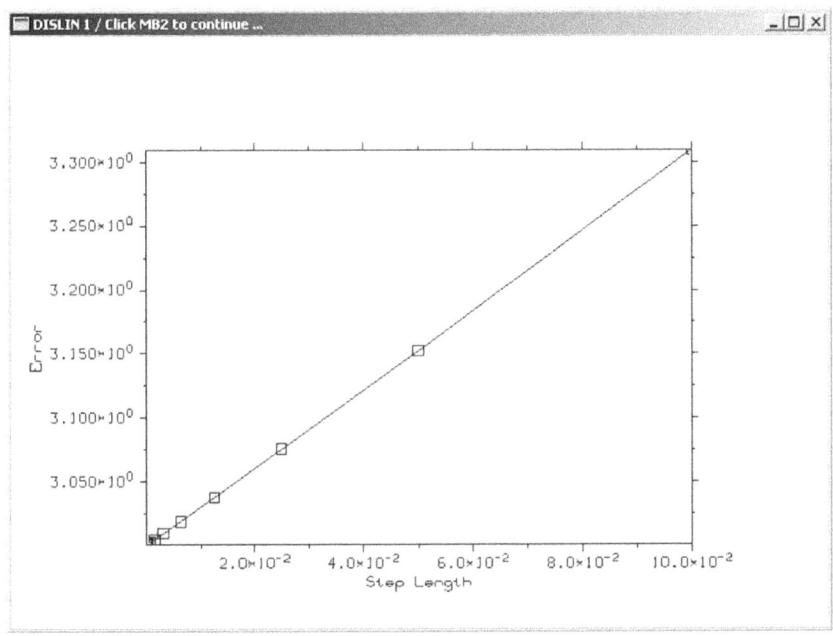

Figure 19.1

```
name( "Step Length", "x" );     // x label
name( "Error", "y" );           // y label
labels( "EXP", "xy" );          // exponential format
incmrk( 1 );                    // markers at every (1) point
setscl( x, 10, "X" );           // automatically scales x axis
setscl( y, 10, "Y" );
//scale( "LOG", "XY" );         // uncomment for log-log plot
float minX, maxX, minY, maxY, stepX, stepY;
graf( minX, maxX, minX, stepX,
      minY, maxY, minY, stepY ); // axes
curve( x, y, 10 );              // plot curve
disfin( );                      // terminate plot
}
```

which yields Figure 19.1, verifying the linear dependence of the error on step length without a mathematical analysis. Similarly, the result $R(\Delta x)$ of an $O(\Delta x)^\alpha$ numerical procedure plotted as a function of $(\Delta x)^\alpha$ yields a straight line for small Δx whose y-intercept is the corrected value. Alternatively, if the result $R_{\Delta x \to 0}$ can be estimated or extrapolated, the associated error estimate varies as $O(\Delta x)^\alpha = R(\Delta x) - R_{\Delta x \to 0}$ for $\Delta x \to 0$. Taking the logarithm of both sides yields $\log(\Delta x)^\alpha = \alpha \log(\Delta x) = \log(R(\Delta x) - R_{\Delta x \to 0})$ so that α corresponds to the slope of the logarithm of this estimated error plotted against $\log(\Delta x)$ (the slope of a log–log plot).

An Octave implementation of the previous program, which omits several specialized features of the DISLIN graph, follows:

```
File: cube.m
function output = cube( aInput )
output = aInput^3;
```

```
File: derivOperator.m
function output = derivOperator( aF, aXValue, aDeltaX )
output = ( aF( aXValue + aDeltaX ) - aF( aXValue ) ) / aDeltaX;
```

```
File: derivplot.m
deltaX = 1.0e-1;
xValue = 1.0;
for loop = 1 : 10
    y( loop ) = derivOperator( @cube, xValue, deltaX );
    x( loop ) = deltaX;
    deltaX = deltaX / 2;
end
plot( x, y, '-s' )
axis tight;
xlabel( 'Step Length' );
ylabel( 'Error' );
```

19.4   Analytic error analysis – higher-order methods

An $O(\Delta x)^2$ accurate finite-difference operator approximation to the continuous derivative can be derived from the Taylor-series relationships

$$f(x + \Delta x) = f(x) + \Delta x \frac{df}{dx} + \frac{(\Delta x)^2}{2!} \frac{d^2 f}{dx^2} + \frac{(\Delta x)^3}{3!} \frac{d^3 f}{dx^3} + \cdots$$

$$f(x - \Delta x) = f(x) - \Delta x \frac{df}{dx} + \frac{(\Delta x)^2}{2!} \frac{d^2 f}{dx^2} - \frac{(\Delta x)^3}{3!} \frac{d^3 f}{dx^3} + \cdots$$

(19.6)

Subtracting the second of these formulas from the first yields the *centered finite-difference operator*

$$\frac{f(x + \Delta x) - f(x - \Delta x)}{2 \Delta x} = \frac{df}{dx} + O(\Delta x)^2$$

(19.7)

In a similar fashion, it can be demonstrated that a fourth-order accurate expression for the derivative is

$$\frac{df}{dx} = \frac{f(x - 2\Delta x) - 8f(x - \Delta x) + 8f(x + \Delta x) - f(x + 2\Delta x)}{12 \Delta x} + O(\Delta x)^4$$

(19.8)

Accordingly, replacing the derivative function in the previous C++ program by

```
double derivOperator(double aF(double),
    double aXValue, double aDeltaX) {
  return ( aF( aXValue + aDeltaX ) - aF( aXValue - aDeltaX ) ) /
    ( 2 * aDeltaX );   // Centered difference approximation
}
```

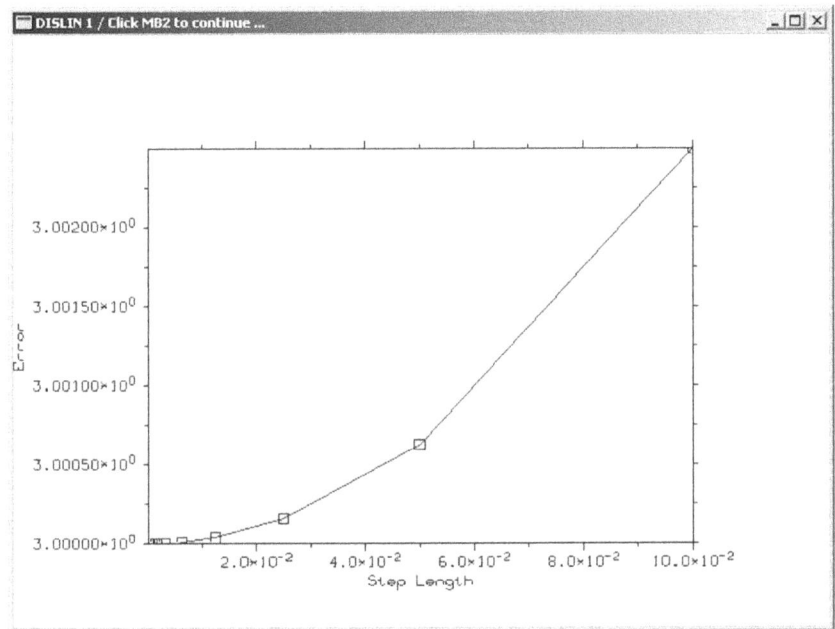

Figure 19.2

and repeating the calculation of Figure 19.1 with Eq. (19.7) yields Figure 19.2. The dependence of the reduced numerical error on $(\Delta x)^2$ is apparent.

19.5 Extrapolation

As an alternative to graphically extrapolating results for different Δx, if the order of accuracy of an arbitrary numerical procedure is known, the results of the method for different step sizes can be combined algebraically. For example, if the result of a calculation with a step length Δx with an error term of the form $O(\Delta x) = a\,\Delta x + O(\Delta x)^2$, where a represents the Δx Taylor-series coefficient in $O(\Delta x)$, is added to the result for the same calculation but with a step length $-\Delta x$, the combined error is $a\,\Delta x + O(\Delta x)^2 + (-a\,\Delta x + O(-\Delta x)^2) = \tilde{O}(\Delta x)^2$. By similarly combining results for step lengths of $\pm\Delta x$, $\pm 2\,\Delta x$, ..., the accuracy of any numerical method can generally be greatly increased if the output varies sufficiently smoothly as a function of x.

19.6 The derivative calculator class

An object-oriented implementation of the derivative operator can be obtained by abstracting a handheld calculator in which the user selects a function to be differentiated, a step length and an evaluation position. The calculator, which stores these quantities, computes the value of the derivative when its calculate button is depressed:

```
class DerivativeCalculator {
   public:
   void setDx( double aDx ) { iDx = aDx; }
   void setX( double aX ) { iX = aX; }
   double dx( ) const { return iDx; }
   double calculateDerivative( ) const
       { return ( iF( iX + iDx ) - iF( iX ) ) / iDx; }
   DerivativeCalculator(double aX, double aDx, double aF( double )) :
       iX( aX ), iDx( aDx ) {
           iF = aF;
   }
   private:
   // Store function address as an internal variable
   double (*iF) ( double );
   double iDx;
   double iX;
};

main ( ) {
   double deltaX = 1.e-1;
   double xValue = 1.0;
   int choice;
   double (*myFunction) ( double );
   cout << "Choose a function 1 - cube, 2 - square "<< endl;
   cin >> choice;
   switch ( choice ) {
       case 1: myFunction = cube; break;      // Output: 3.31
       case 2: myFunction = linear; break;    // Output: 1
       default: cout << "Incorrect Input - program
           exiting";
   }
   DerivativeCalculator DC1( 1.0, 0.1, myFunction );
   cout << DC1.calculateDerivative( );
}
```

19.7 Integration

The definite integral

$$I^L f(x) = \int_L^x f(x')dx' \tag{19.9}$$

is again an operator that transforms the integrand function $f(x)$ into a second function $I^L f(x)$. This continuous operator is again implemented numerically by referring to its underlying definition as a limit of a discrete expression,

$$I^L f(x) = \lim_{n \to \infty} I_n^L f(x) = \lim_{n \to \infty} \sum_{k=0}^{n-1} \Delta x \, a_k \tag{19.10}$$

in which $\Delta x = (x - L)/N$ and the a_k are suitably chosen values of $f(x)$ within the interval $[x + k\,\Delta x, \; x + (k+1)\Delta x]$. The summation limits from 0 to $n-1$ yield n intervals of length Δx. A common, subtle programming error is to set the upper limit to n, yielding an $O(\Delta x)$ numerical error that is often mistaken for the error of the numerical method.

In calculus, two possible choices for the a_k are commonly cited. The *rectangular rule*,

$$a_k = f(L + k\,\Delta x) \tag{19.11}$$

evaluates the function at the left endpoint of each interval, while the *midpoint rule*

$$a_k = f(L + (k + 0.5)\Delta x) \tag{19.12}$$

instead employs the interval midpoint. The midpoint rule Octave program

File myMidpoint.m:

```
function result = myMidpoint( aFunction, leftEndPoint,  ...
            rightEndPoint, numberOfIntervals )
deltaX = ( rightEndPoint - leftEndPoint ) / numberOfIntervals;
result = deltaX * sum( aFunction( leftEndPoint + 0.5 * deltaX + ...
            deltaX * ( 0 : numberOfIntervals - 1 ) ) )
```

is called for a sine function argument as follows:

```
myMidpoint( @sin, 0., pi, 40 )
```

While the midpoint and rectangular rules yield identical results in the continuous limit ($\Delta x \to 0$), the discrete errors differ. For example, the difference between the rectangular-rule approximation and the exact result equals

$$E_{\text{rect}} = \sum_{k=0}^{n-1} \int_{x_k}^{x_k + \Delta x} [f(x_k) - f(x)]dx \tag{19.13}$$

If $f(x)$ is continuous in the interval from L to R, it can be expanded in the kth interval in a Taylor series in $(x_k - x)$, yielding

$$E_{\text{rect}} = \sum_{k=0}^{n-1} \int_{x_k}^{x_k + \Delta x} f'(x_k)(x_k - x)dx + O(\Delta x)^2 \tag{19.14}$$

However, if the magnitude of the first derivative of $f(x)$ is bounded on the interval $x \in [L, R]$ by M_1, then over the entire interval

$$|E_{\text{rect}}| < \frac{M_1 n (\Delta x)^2}{2} = \frac{M_1 (R - L)\Delta x}{2} \tag{19.15}$$

since $n\,\Delta x$ is the total length of the integration interval. Therefore, unless there occurs an exceptional cancellation between positive and negative contributions to the total error, the error varies linearly with Δx since the error in each discrete interval is $O(\Delta x)^2$ but the errors add over $L/\Delta x$ intervals. The overall error, however, decreases to $O(\Delta x)^2$ with the midpoint rule, as can be demonstrated by extending the above methodology.

The integration accuracy can be further increased by employing Simpson's rule:

$$\int_L^R f(x)dx = \frac{\Delta x}{6} \sum_{i=0}^{n-1} \left[f(x_i) + 4f\left(\frac{x_i + x_{i+1}}{2}\right) + f(x_{i+1}) \right] + O(x_{i+1} - x_i)^4 \quad (19.16)$$

The following Octave program implements Simpson's rule over an interval from **leftEndPoint** to **rightEndPoint**, noting that the values of $f(x_1), f(x_2), \ldots, f(x_{n-1})$ are evaluated twice in the above expression, yielding weights of $1, 4, 2, 4, 2, \ldots,$ $4, 1$:

```
function result = mySimpson( aFunction, aLeftEndPoint, ...
            aRightEndPoint, aNumberOfIntervals )
deltaX = ( aRightEndPoint - aLeftEndPoint ) / aNumberOfIntervals;
xLeft = leftEndPoint
deltaX2 = deltaX / 2;
mySum = 0;
for loop = 0 : aNumberOfIntervals - 1
        mySum = mySum + 2 * aFunction( xLeft ) + ...
                4 * aFunction( xLeft + deltaX2 );
        xLeft = xLeft + deltaX;
end
mySum = mySum + aFunction( aRightEndPoint ) ...
            - aFunction( aLeftEndPoint );
result = deltaX * mySum / 6;
```

19.8 Root-finding procedures

The roots of a real equation $f(x) = 0$ over an interval $[L, R]$ are most reliably obtained by the bisection method, which requires only the continuity of $f(x)$. First $f(x)$ is evaluated on a sufficient number of equally spaced points over the interval to insure that no more than one root occurs between each set of adjacent points. Each interval within which the function changes sign is then considered in turn. Labeling the current interval $[a, b]$ and its bisector c, if $f(b)f(c) \leq 0$ the root falls between b and c including possibly the endpoints of the interval. Accordingly, the current interval is replaced by $[b, c]$; otherwise, it is set to $[a, b]$. The process is continued until the endpoint changes by less than a prescribed value, the *error* over one iteration step:

```
function result = myBisect( aFunction, aLeftLimit, ...
            aRightLimit, aError )
while ( abs( aLeftLimit - aRightLimit ) > aError )
        midpoint = ( aLeftLimit + aRightLimit ) / 2;
        if aFunction ( midpoint ) * aFunction( aRightLimit ) <= 0
            aLeftLimit = midpoint;
        else
```

```
                    aRightLimit = midpoint;
         end
end
result = ( aLeftLimit + aRightLimit ) / 2;
```

The bisection method is applicable to any continuous function, but lacks computational efficiency, since each iteration only halves the extent of the interval containing the root. *Newton's method*, in contrast, converges rapidly but fails if a function is not differentiable or possesses a very small slope near its root, or for an inaccurate initial estimate, x_1, of the root. The procedure locates the root of $f(x)$ near x_1, by approximating $f(x)$ by the first two terms in its Taylor-series expansion,

$$f_T(x) \approx f(x_1) + \frac{df(x_1)}{dx}(x - x_1) \tag{19.17}$$

The zero of the above linear approximation to $f(x)$,

$$x_2 = x_1 - \frac{f(x_1)}{\left(\dfrac{df(x_1)}{dx}\right)} \tag{19.18}$$

is generally far closer to the exact root than x_1. The minus sign in the above equation is often mistakenly omitted. The derivative of $f(x)$ in Eq. (19.18) is evaluated numerically or analytically and the procedure iterated until $|x_{i+1} - x_i| < \varepsilon$. The above formula can also be derived geometrically by noting that, from the equal ratios of equivalent sides of similar triangles, $\Delta f(x_1)/\Delta x \approx (f(x_1) - f(x_2))/(x_1 - x_2)$ with $f(x_2) = 0$.

A non-recursive implementation of Newton's method in Octave is given by

```
function result = newton( aFunction, aEstimate, aError )
deltaX = 1.e-3;
for loop = 1 : 5
        derivative = ( aFunction( aEstimate + deltaX ) - ...
                aFunction( aEstimate - deltaX ) ) / ( 2. * deltaX );
        deltaX = -aFunction( aEstimate ) / derivative;
        aEstimate = aEstimate + deltaX;
        aFunction( aEstimate )
        if abs( deltaX ) < aError
                break;
        end
end
result = aEstimate
```

A recursive Java implementation of Newton's method, in which the technique of Section 18.3 is employed to select $f(x)$ at runtime follows, where **System.in.read()** returns a single character extracted from the terminal as an **int**:

```
interface MyFunction {
        double myFunction( double aD );
}
```

```
public class Newton {
  static int count;
  static double derivative( MyFunction aMF, double aX ) {
    double deltaX = 1.e-3;
    return ( aMF.myFunction( aX + deltaX )
      - aMF.myFunction( aX - deltaX ) ) / ( 2. * deltaX );
  }
  static double newton( MyFunction aMF1, double aEstimate,
    double aError ){
    double deltaX = - aMF1.myFunction( aEstimate )
      / derivative( aMF1, aEstimate );
    aEstimate += deltaX;
    if ( Math.abs( deltaX ) < aError || count++ == 40 )
      return aEstimate;
    else return newton( aMF1, aEstimate, aError );
  }
  public static void main ( String [ ] args ) throws Exception {
    System.out.println( "Insert 1 for square, 2 for cube" );
    char c = (char) System.in.read( );
    MyFunction MF1;
    switch ( c ) {
      case '1': MF1 = new MyFunction( ) { public double
        myFunction( double aD ) { return aD * aD - 4; } } ;
        break;
      case '2': MF1 = new MyFunction( ) { public double
      myFunction( double aD ) { return aD * aD * aD - 27; } } ;
      break;
      default: System.out.println( "Error in input" ); return;
      }
    double estimate = 2.5;
    double error = 1.e-8;
    System.out.println( newton( MF1, estimate, error ) );
  }
}
```

Newton's method can converge slowly or even diverge unless the derivative is evaluated analytically. Otherwise, once the numerical estimate is close to the root, the $O(\Delta x)^2$ error in the central difference operator imparts an error of the same magnitude to the root value. While this problem can be alleviated with a high-order approximation for the derivative and a small **delta**, truncation errors then increase, especially if $f(x)$ varies slowly near the root. Further, if x_1 is not sufficiently near the root, the sign of df/dx can be the opposite of its sign at the root. Hence, x_2 is positioned further from the root than x_1, and successive iterations generally diverge. An interval $[a, b]$ containing the root can then be selected with x_1 set to a. If x_2 falls within the interval, Newton's method is applied; otherwise, the bisection method is employed to generate an improved starting value.

19.9 Minimization

While the minimum or maximum values of a function can be obtained from the roots of its numerical derivative, approximating the derivative operator yields numerical errors. Instead, to implement a highly simplified minimization routine, the function is first evaluated at three equidistant points (l, m, r) within an interval $[l, r]$ that contains the minimum. Subsequently, if $f(r) > f(l)$, $[m, r]$ is bisected by a point x. If $f(x) < f(m)$, the new interval endpoints are chosen as $[m, r]$, otherwise the endpoints are set to $[l, x]$. This yields (in Octave)

```
function result = myMinimum( aFunction, aLeftEndPoint, ...
      aRightEndPoint, aError );
leftValue = aFunction( aLeftEndPoint );
rightValue = aFunction( aRightEndPoint );
for loop = 1 : 500
      middlePoint = ( aLeftEndPoint + aRightEndPoint ) / 2;
      Middlevalue = aFunction( middlePoint );
          aNewPoint = ( aRightEndPoint + middlePoint ) / 2;
          newValue = aFunction( aNewPoint );
          if ( newValue < middleValue )
              aLeftEndPoint = middlePoint;
      else
              aRightEndPoint = aNewPoint;
      end
      else
          aNewPoint = ( aLeftEndPoint + middlePoint ) / 2;
          newValue = aFunction( aNewPoint );
          if ( newValue < middleValue )
              aRightEndPoint = middlePoint;
      else
              aLeftEndPoint = aNewPoint;
      end
      if aRightEndPoint - aLeftEndPoint < aError
              break;
      end
end
result = middlePoint;
```

The minimum of the parabola passing through the function values at the three evaluation points can also be employed to improve significantly the choice of points in the successive iteration step.

Chapter 20
Linear algebra

Matrices typically describe systems of coupled components, generally subject to external or internal constraints. This chapter considers the standard numerical procedures for such problems, namely linear-equation and eigenvalue solvers. Since matrix calculations are often resource-intensive and error-prone, favoring optimized library routines, the discussion focuses on basic principles.

20.1 Matrices

A linear transformation of variables, or equivalently a linear system of equations, can be represented by a matrix. For example, the Lorenz transformation from a system to a second system traveling at a relative velocity v, namely

$$
\begin{aligned}
ct' &= \gamma(ct - \beta x) \\
x' &= \gamma(x - \beta ct) \\
y' &= y \\
z' &= z
\end{aligned}
\tag{20.1}
$$

with $\beta = v/c$ and $\gamma = 1/\sqrt{1 - v^2/c^2}$ is linear since scaling all the input variables scales the output by the same factor. The transformation is equivalently expressed as

$$
\begin{pmatrix} ct' \\ x' \\ y' \\ z' \end{pmatrix} = \begin{pmatrix} \gamma & -\beta\gamma & 0 & 0 \\ -\beta\gamma & \gamma & 0 & 0 \\ 0 & 0 & 1 & 0 \\ 0 & 0 & 0 & 1 \end{pmatrix} \begin{pmatrix} ct \\ x \\ y \\ z \end{pmatrix}
\tag{20.2}
$$

A second transformation applied to the primed variables leads to matrix multiplication when referred to the unprimed variables.

20.2 Linear-equation solvers

Since the direct problem associated with Eq. (20.1) or Eq. (20.2) expresses the transformed variables by multiplication and addition of the untransformed coordinates, the inverse problem of finding the untransformed from the transformed coordinates generalizes the operation of division and is therefore more involved. The Gaussian elimination procedure for determining the c_i in the system of N linear equations

$$\sum_{j=0}^{N-1} A_{ij}c_j = b_i \tag{20.3}$$

or equivalently $\mathbf{c}$ in the matrix equation system $\mathbf{Ac} = \mathbf{b}$, where the elements of the $N \times N$ matrix $\mathbf{A}$ are A_{ij} and the elements of the vector $\mathbf{b}$, b_i, are given, repeatedly scales and subtracts matrix rows from other rows to recast the original set of equations into tridiagonal form. The final equation in this new set possesses the form $A_{N-1,N-1}^{(N-1)} c_{N-1} = b_{N-1}^{(N-1)}$, yielding c_{N-1}. Subsequently, through "back substitution", the preceding equation, which contains only c_{N-1}, and c_{N-2} is solved for c_{N-2} and so on.

The following simple example,

$$\begin{aligned} c_0 + 2c_1 &= 1 \\ 3c_0 + c_1 &= 2 \end{aligned} \tag{20.4}$$

illustrates the procedure. Multiplying the first equation in this set by 3 and subtracting it from the second equation yields the tridiagonal system

$$\begin{aligned} c_0 + 2c_1 &= 1 \\ -5c_1 &= -1 \end{aligned} \tag{20.5}$$

In the back-substitution step, the second of these equations is solved for c_1 and the result inserted into the first equation to obtain c_0.

By analogy, subtracting A_{i0}/A_{00} times the first row of Eq. (20.3) from each row $i \neq 0$ generates a new equation system for the c_j with elements

$$A_{ij}^{(1)} = A_{ij} - A_{0j}\frac{A_{i0}}{A_{00}}, \qquad b_i^{(1)} = b_i - b_0\frac{A_{i0}}{A_{00}} \tag{20.6}$$

Here the first element is absent from each row except the first ($i = 0$) row. This procedure is then repeated within the submatrix formed by excluding the first row and column from the matrix system. After $N-1$ iterations, the tridiagonal equation system

$$\begin{aligned} A_{00}c_0 + A_{01}c_1 + \cdots + A_{0N-1}c_{N-1} &= b_0 \\ A_{11}^{(1)}c_1 + \cdots + A_{1N-1}^{(1)}c_{N-1} &= b_1^{(1)} \\ &\vdots \\ A_{N-1,N-1}^{(N-1)}c_{N-1} &= b_{N-1}^{(N-1)} \end{aligned} \tag{20.7}$$

is obtained. To implement back substitution the last, trivial, equation is solved and the result for c_{N-1} is inserted in the previous equation; subsequently the values for both c_{N-1} and c_{N-2} are inserted in the third from the last equation and the process is repeated until the first equation is reached. This yields

$$c_{N-1} = b_{N-1}^{(N-1)} \Big/ A_{N-1,N-1}^{(N-1)}$$

$$c_{N-2} = \left(b_{N-2}^{(N-2)} - A_{N-2,N-1}^{(N-2)} c_{N-1}\right) \Big/ A_{N-2,N-2}^{(N-2)} \tag{20.8}$$

$$c_{N-3} = \left(b_{N-3}^{(N-3)} - A_{N-3,N-2}^{(N-3)} c_{N-2} - A_{N-3,N-1}^{(N-3)} c_{N-1}\right) \Big/ A_{N-3,N-3}^{(N-3)}$$

With the above notation, Gaussian elimination is coded in C++ as (where the input two- and one- dimensional array parameters **aA** and **aB** are overwritten)

```cpp
#include <stdio.h>
const int n = 2;
// Note: both aA and aB are overwritten
void gauss( double aA[ ][n], double aC[ ], double aB[ ] ) {
      // Forward elimination
      for ( int i = 0; i < n; i ++ ) {
            if ( !aA[i][i] ) exit( 0 );
            for ( int j = i + 1; j < n; j++ ) {
                  double d = aA[j][i] / aA[i][i];
                  for ( int k = i + 1; k < n; k++ )
                        aA[j][k] -= d*aA[i][k];
                  aB[j] -= d*aB[i];
            }
      }
      if ( !aA[n-1][n-1] ) exit( 0 );
      // Back substitution
      for ( int i = n - 1; i >= 0; i-- ) {
            aC[i] = aB[i];
            for ( int j = i + 1; j < n; j++ )
                  aC[i] -= aA[i][j] * aC[j];
            aC[i] /= aA[i][i];
      }
}
main( ) {
      double a[n][n] = { { 1, 2 }, { 3, 8 } };
      double b[n] = { 2, 5 };
      double c[n];
      gauss( a, c, b );
      cout << c[0] << '\t' << c[1] << endl;
}
```

Since the second derivative operator is represented by the three-point formula of Eq. (20.8), the specialization of the above routine to tridiagonal matrices is employed in solving, for example, diffusion and wave equations. To preserve memory space, the diagonal and the upper and lower co-diagonals of the matrices are typically stored as three separate arrays of dimensions N, $N-1$ and $N-1$, respectively. A C++ tridiagonal equation solver that preserves the input parameters is

```cpp
void tridiagonalSolver( double aLowerCodiagonal[ ],
      double aDiagonal[ ], double aUpperCodiagonal[ ],
      double aInputVector[ ], double aOutputVector[ ],
      double aScratch[ ], int aNumberOfPoints ) {
      // Forward elimination
```

```
        double bsave = aDiagonal[0];
        if ( !bsave ) exit( 0 );
        aOutputVector[0] = aInputVector[0] / bsave;
        for ( int loop = 1; loop < aNumberOfPoints; loop++ ) {
                aScratch[loop] = aUpperCodiagonal[loop - 1] / bsave;
                bsave = aDiagonal[loop] - aScratch[loop] *
                    aLowerCodiagonal[loop - 1];
        if ( !bsave ) exit( 0 );
        // Back substitution
        aOutputVector[loop] = ( aInputVector[loop] -
            aLowerCodiagonal[loop - 1] * aOutputVector[loop - 1] )
            / bsave;
        }
        for ( int loop = aNumberOfPoints - 2; loop > -1; loop-- )
            aOutputVector[loop] -= aScratch[loop + 1]
            * aOutputVector[loop + 1];
}

main( ) {
        double diagonal[2] = { 1, 8 };
        double upperCodiagonal[1] = { 2 };
        double lowerCodiagonal[1] = { 3 };
        double inputVector[2] = { 2, 5 };
        double outputVector[2];
        double scratch[2];
        tridiagonalSolver( lowerCodiagonal, diagonal, upperCodiagonal,
                inputVector, outputVector, scratch, 2 );
        cout << outputVector[0] << '\t' << outputVector[1] << endl;
}
```

Often the check for non-zero components of **bsave** can be omitted. If A_{ij} remains unchanged over multiple realizations, the above codes can be accordingly modified to improve efficiency.

In Octave, function parameters are passed by value so that additional (scratch) space for preserving the function arguments is superfluous. A tridiagonal matrix program for either column and row vectors that does not examine **bsave** can be coded as

```
function outputVector = myTridigonal( aLowerCodiagonal, ...
            aDiagonal, aUpperCodiagonal, aInputVector )
numberOfEquations = length( aDiagonal );
outputVector = zeros( 1, numberOfEquations );
% Forward elimination
for loop = 2 : numberOfEquations
        temporary = aLowerCodiagonal(loop - 1) / aDiagonal(loop - 1);
        aDiagonal(loop) = aDiagonal(loop) ...
            - temporary * aUpperCodiagonal(loop - 1);
        aInput(loop) = aInputVector(loop) ...
            - temporary * aInput(loop - 1);
end
% Back substitution
outputVector(numberOfEquations) = aInput(numberOfEquations) ...
            / aDiagonal(numberOfEquations);
for loop = numberOfEquations - 1 : -1 : 1
```

```
outputVector(loop) = ( aInput(loop) - aUpperCodiagonal(loop)  ...
            * outputVector(loop + 1) ) / aDiagonal(loop);
end
```

20.3   Errors and condition numbers

Numerical errors in Gaussian elimination principally arise from two sources. The first of these occurs when, for example, $|A_{00}| \ll |A_{0i}|$ in Eq. (20.6) so that typically $A_{0j} A_{i0} / A_{00} \gg A_{ij}$. In this case, the number of significant digits retained from A_{ij} is reduced after the subtraction. This error, which can appear at each forward step of the algorithm, can be eliminated by pivoting. This refers to interchanging the order of rows in the equation system at each forward step to minimize the magnitude of the subtracted values.

A second, more problematic source of error results if the equation system is nearly linearly dependent; that is, if multiplying a number of equations by different coefficients and summing approximates a different equation in the set. Small changes in the input vector, **b**, can then induce large variations in the solution vector **c**. Geometrically, for $N = 2$, two nearly identical equations describe almost parallel lines. A small change in the equation of one line resulting from a minor variation in **b** substantially displaces their point of intersection and therefore leads to a markedly different solution vector **c**. In N dimensions, a solution occurs at the point of intersection of N hyperplanes with $N - 1$ dimensions, and therefore behaves similarly.

The deviation of an equation system from a linearly dependent system is quantified by the *condition number* of the corresponding matrix, which is defined as the ratio of the largest to the smallest of the matrix eigenvalues. A singular matrix possesses an infinite condition number, whereas the condition number of an "ill-conditioned" matrix exceeds 10^6 in single precision and 10^{12} in double precision. Routines for computing condition numbers are present in most numerical libraries since they are frequently incorporated into matrix solvers.

20.4   Application: least-squares procedure

Least-squares fitting provides an important application for equation-solution methods. The *linear* least-squares procedure estimates the parameters a_i of a model of the form

$$y(x, a_1, \ldots, a_M) = \sum_{j=1}^{M} a_j Y_j(x) \tag{20.9}$$

in which the $Y_i(x)$ are given functions, from noisy experimental data by minimizing the deviation of the model predictions from the data.

Given a set of N measurement points (x_i, y_i), if the random measurement error at each x_i is known beforehand to be σ_i and the actual physical value of the output variable is $y_{\text{exact}}(x_i)$, the probability of observing the measured value y_i for errors distributed according to a Gaussian (normal) probability distribution is given by

$$p(y_i) = \frac{1}{\sigma\sqrt{\pi}} e^{-\frac{(y_i - y_{\text{exact}}(x_i))^2}{2\sigma_i^2}} \tag{20.10}$$

If the errors at each x_i are mutually uncorrelated, the probability of observing a certain set of data points is the product of the individual probabilities for each point:

$$p(\{y_i\}) \propto \prod_{i=1}^{N} e^{-\frac{(y_i - y_{\text{exact}}(x_i))^2}{2\sigma_i^2}} = e^{-\sum_{i=1}^{N} \frac{(Y(x_i) - y_i)^2}{\sigma_i^2}} \equiv e^{-\chi^2} \tag{20.11}$$

Accordingly, if y_{exact} is approximated by $y(x, a_1, \ldots, a_N)$, the optimal parameters a_i minimize the *chi-squared function*

$$\chi^2(a_1, a_2, \ldots, a_N) \equiv \sum_{i=1}^{N} \frac{(Y(x_i) - y_i)^2}{\sigma_i^2} \tag{20.12}$$

which requires

$$\frac{\partial \chi^2}{\partial a_k} = \frac{\partial}{\partial a_k} \sum_{i=1}^{N} \frac{1}{\sigma_i^2} \left\{ \sum_{j=1}^{M} a_j Y_j(x_i) - y_i \right\}^2$$

$$= 2 \sum_{i=1}^{N} \frac{1}{\sigma_i^2} \left[\frac{\partial}{\partial a_k} \left\{ \sum_{j=1}^{M} a_j Y_j(x_i) - y_i \right\} \left(\sum_{j=1}^{M} a_j Y_j(x_i) - y_i \right) \right]$$

$$= 0 \tag{20.13}$$

for each i. This yields the M linear equations

$$\sum_{i=1}^{N} \frac{Y_k(x_i)}{\sigma_i} \left(\sum_{j=1}^{M} a_j \frac{Y_j(x_i)}{\sigma_i} - \frac{y_i}{\sigma_i} \right) = 0 \tag{20.14}$$

or, in terms of the $N \times M$ matrix $\mathbf{A}$ with $A_{ij} = Y_j(x_i)/\sigma_i$ and the vector $\mathbf{b}$ with $b_i = y_i/\sigma_i$,

$$\mathbf{Ma} \equiv \left(\mathbf{A}^{\mathsf{T}} \mathbf{A} \right) \mathbf{a} = \mathbf{A}^{\mathsf{T}} \mathbf{b} \tag{20.15}$$

If all measurements possess the same error, Eq. (20.15) becomes $\tilde{\mathbf{M}}\mathbf{a} = \mathbf{d}$, with

$$\tilde{M}_{kj} = \sum_{i=1}^{N} Y_k(x_i) Y_j(x_i), \qquad d_k = \sum_{i=1}^{N} Y_k(x_i) y_i \tag{20.16}$$

If there exist fewer x_i values than parameters in the model function, so that $N < M$, the equation system of Eq. (20.15) is *underdetermined*. The *singular value decomposition* (SVD) method is then employed to obtain an approximate solution.

The least-squares procedure simplifies considerably when fitting a line $y = a_1 + a_2 x$ to a set of data points for which the equations for a_1 and a_2 are given by

$$\frac{\partial}{\partial a_1} \sum_{i=1}^{N} \frac{1}{\sigma_i^2} (a_1 + a_2 x_i - y_i)^2 = 2 \sum_{i=1}^{N} \frac{1}{\sigma_i^2} (a_1 + a_2 x_i - y_i) = 0$$

$$\frac{\partial}{\partial a_2} \sum_{i=1}^{N} \frac{1}{\sigma_i^2} (a_1 + a_2 x_i - y_i)^2 = 2 \sum_{i=1}^{N} \frac{x_i}{\sigma_i^2} (a_1 + a_2 x_i - y_i) = 0$$

(20.17)

which yields

$$a_1 \sum_{i=1}^{N} \frac{1}{\sigma_i^2} + a_2 \sum_{i=1}^{N} \frac{x_i}{\sigma_i^2} - \sum_{i=1}^{N} \frac{y_i}{\sigma_i^2} = 0$$

$$a_1 \sum_{i=1}^{N} \frac{x_i}{\sigma_i^2} + a_2 \sum_{i=1}^{N} \frac{x_i^2}{\sigma_i^2} - \sum_{i=1}^{N} \frac{x_i y_i}{\sigma_i^2} = 0$$

(20.18)

and therefore

$$a_1 = \frac{\sum_{i=1}^{N} \frac{y_i}{\sigma_i^2} \sum_{i=1}^{N} \frac{x_i^2}{\sigma_i^2} - \sum_{i=1}^{N} \frac{x_i}{\sigma_i^2} \sum_{i=1}^{N} \frac{x_i y_i}{\sigma_i^2}}{\sum_{i=1}^{N} \frac{1}{\sigma_i^2} \sum_{i=1}^{N} \frac{x_i^2}{\sigma_i^2} - \left(\sum_{i=1}^{N} \frac{x_i}{\sigma_i^2} \right)^2}, \quad a_2 = \frac{\sum_{i=1}^{N} \frac{1}{\sigma_i^2} \sum_{i=1}^{N} \frac{x_i y_i}{\sigma_i^2} - \sum_{i=1}^{N} \frac{x_i}{\sigma_i^2} \sum_{i=1}^{N} \frac{y_i}{\sigma_i^2}}{\sum_{i=1}^{N} \frac{1}{\sigma_i^2} \sum_{i=1}^{N} \frac{x_i^2}{\sigma_i^2} - \left(\sum_{i=1}^{N} \frac{x_i}{\sigma_i^2} \right)^2}$$

(20.19)

20.5   Eigenvalues and iterative eigenvalue solvers

An eigenvector x of $\mathbf{A}$ solves $\mathbf{A}x = \lambda_i x$ in which the constant λ_i is termed an eigenvalue. If $\mathbf{A}$ is an $N \times N$ Hermitian matrix, it possesses N linearly independent eigenvectors with real eigenvalues such that any N-component vector $\mathbf{x}$ can be expressed as a linear combination of these eigenvectors. To find iteratively the eigenvector, here assumed unique, with eigenvalue closest to an initial estimate $\lambda^{(0)}$ the equation system

$$(\mathbf{A} - \lambda^{(i)} \mathbf{I}) \mathbf{x}^{(i+1)} = \mathbf{x}^{(i)}$$

(20.20)

must be solved repeatedly. Here $\mathbf{I}$ is the identity matrix and $\mathbf{x}^{(i)}$ is the ith estimate for the corresponding eigenvector, where the initial estimate, $\mathbf{x}^{(0)}$ can be chosen effectively randomly. Expressing $\mathbf{x}^{(i)}$ as a linear combinations of the eigenvectors, Θ_k, of $\mathbf{A}$,

$$\mathbf{x}^{(i)} = \sum_{k=1}^{N} c_k^{(i)} \Theta_k$$

(20.21)

the recursion relation yields

$$\sum_{k=1}^{N} (\lambda_k - \lambda^{(i)}) c_k^{(i+1)} \Theta_k = \sum_{k=1}^{N} c_k^{(i)} \Theta_k \tag{20.22}$$

However, since the eigenvectors, Θ_k, are linearly independent, the coefficients of each eigenvector on both sides of the equation must be separately equal. Therefore,

$$c_k^{(i+1)} = \frac{1}{\lambda_k - \lambda^{(i)}} c_k^{(i)} \tag{20.23}$$

enhancing the relative amplitude of the eigenvector in $\mathbf{c}^{(i+1)}$ closest to $\lambda^{(i)}$.

While $\lambda^{(i)}$ can remain equal to $\lambda^{(0)}$, $x^{(i+1)}$ yields an improved value of this parameter. On multiplying both sides of Eq. (20.20) by the transpose of the column vector $\mathbf{x}^{(i+1)}$, we obtain

$$(\mathbf{x}^{(i+1)})^{\mathrm{T}} \mathbf{A} \mathbf{x}^{(i+1)} - \lambda^{(i)} (\mathbf{x}^{(i+1)})^{\mathrm{T}} \mathbf{x}^{(i+1)} = (\mathbf{x}^{(i+1)})^{\mathrm{T}} \mathbf{x}^{(i)} \tag{20.24}$$

Since $\mathbf{x}^{(i+1)}$ approaches the eigenvector with eigenvalue nearest $\lambda^{(i)}$, an improved eigenvalue estimate is obtained from $(\mathbf{x}^{(i+1)})^{\mathrm{T}} \mathbf{A} \mathbf{x}^{(i+1)} \approx \lambda^{(i+1)} (\mathbf{x}^{(i+1)})^{\mathrm{T}} \mathbf{x}^{(i+1)}$ or

$$\lambda^{(i+1)} = \lambda^{(i)} + \frac{(\mathbf{x}^{(i+1)})^{\mathrm{T}} \mathbf{x}^{(i)}}{(\mathbf{x}^{(i+1)})^{\mathrm{T}} \mathbf{x}^{(i+1)}} \tag{20.25}$$

The amplitude of $\mathbf{x}^{(i)}$ varies exponentially with i and must be periodically renormalized.

Chapter 21
Fourier transforms

The frequency content of a continuous signal $s(t)$ is uniquely determined by its continuous Fourier transform. However, if the signal is instead sampled at discrete times $t_m = t_0 + m\Delta t, \ m = 0, 1, \ldots, N - 1$, then, although its frequency behavior can be specified in terms of N values, these depend on the assumed behavior of the signal between the sample points and outside the interval. The discrete Fourier transform consequently possesses more involved properties than those of the continuous transform.

The *discrete complex Fourier transform* (DFT) is defined as

$$S(m) = \sum_{l=0}^{N-1} s(t + l\,\Delta t)e^{-i\omega_m l\,\Delta t} = \sum_{l=0}^{N-1} s(t + l\,\Delta t)\cos(\omega_m l\,\Delta t)$$
$$- i \sum_{l=0}^{N-1} s(t + l\,\Delta t)\sin(\omega_m l\,\Delta t) \tag{21.1}$$

The frequencies ω_m are set to the equally spaced values

$$\omega_m = 2\pi m/(N\,\Delta t), \tag{21.2}$$

with $m = 0, 1, \ldots, N - 1$. The real part of $S(m)$ is termed the *cosine transform*, while the imaginary part is the negative *sine transform*. Upon introducing the notation $s_l \equiv s(t + l\,\Delta t)$,

$$S(m) = \sum_{l=0}^{N-1} s_l e^{-i\frac{2\pi ml}{N}} \tag{21.3}$$

for which the inverse transform is

$$s_k = \frac{1}{N} \sum_{m=0}^{N-1} S(m)e^{i\frac{2\pi mk}{N}} \tag{21.4}$$

as can be verified by direct substitution.

For large N, Eq. (21.3) admits rapid evaluation if N possesses many factors of 2. To demonstrate, for $N = 8$, if l and m are both written in binary representation as $l = 4l_3 + 2l_2 + l_1$ and $m = 4m_3 + 2m_2 + m_1$ then the sum in Eq. (21.3) takes the form

$$S(m) = \sum_{l_3=0}^{1} \sum_{l_2=0}^{1} \sum_{l_1=0}^{1} s_{4l_3+2l_2+l_1} e^{-i\frac{2\pi}{8}(4m_3+2m_2+m_1)(4l_3+2l_2+l_1)} \qquad (21.5)$$

For $m = 4$, the terms become

$$e^{-i\frac{2\pi}{8}4(4l_3+2l_2+l_1)} = e^{-i\frac{2\pi}{8}4l_1} \qquad (21.6)$$

reducing the total number of addition operations to 2. In this manner, if $N = 2^M$ the number of "fast Fourier transform" (FFT) operations decreases from $O(N^2)$ to $O(NM)$.

To illustrate some features of the FFT, type into Octave

```
signalR = sin( 2 * pi / 64 * [0 : 63] )
fftSignalR = fft( signalR );
plot( fftSignalR );
```

The graph appears meaningless since the FFT yields a complex result. On the other hand,

```
plot( real( fftSignalR ) );
```

and

```
plot( imag( fftSignalR ) );
```

display the cosine and the negative of the sine transforms. Since the signal $s(m)$ is a sine function, the first of these is zero to within numerical precision, while the negative sine transform of the signal has a negative peak at the second, $m = 1$, point in the computational window, indicating, as expected from Eq. (21.2), that the lowest non-zero frequency is $\omega_1 = 2\pi/(N \Delta t)$ in the DFT or, equivalently, $\Delta f = \omega/(2\pi) = 1/2\pi = 1/(N \Delta t)$. Since the frequency spacing therefore varies with the duration of the signal's time record (the computational window width), to increase the frequency resolution, the window must be broadened, even for localized signals.

Further, since $\sin(\omega t) = (\exp(i\omega t) - \exp(-i\omega t))/(2i) = -i(\exp(i\omega t) - \exp(-i\omega t))/2$, the negative frequency in the sine function yields a value of $+32i$ at the last, 64th point in the computational window (the sine transform evaluated at a negative frequency is the negative of its value for the equivalent positive frequency since the sine function is odd). Since the first point, $m = 1$ in Eq. (21.3), of the DFT corresponds to zero frequency, the $(N/2 + 1)$th point, corresponds to $m = N/2$ and therefore $\omega_m = \pi/\Delta t$, which is termed the *Nyquist frequency*. This constitutes the highest positive frequency (or equivalently the largest negative frequency, since the DFT is periodic in frequency as well as time) in the transform. The (next) largest negative frequency appears at the subsequent point with $m = N/2 + 1$, since

$$e^{-i\frac{2\pi(N/2+1)l}{N}} = e^{-i\left(\frac{2\pi(N/2+1)l}{N} - 2\pi l\right)} = e^{-i\frac{2\pi(N/2+1-N)l}{N}} = e^{-i\frac{2\pi(-(N/2-1))l}{N}} \qquad (21.7)$$

The smallest negative frequency, $\Delta f = -1/(N \Delta t)$, is situated at the last point, $m = N - 1$. To recover the original signal, the inverse FFT can be applied:

```
resultR = ifft( fftSignalR );
plot( real ( resultR ) );
```

Since the FFT remains unchanged if either m or l in Eq. (21.3) is incremented by N, the FFT and inverse FFT exhibit periodicity both in time and in frequency. Effectively, the Fourier transform computes the frequency content as if $s(t)$ were periodically extended outside the data window such that $s(t) = s(t + lN \Delta t)$, where l is any integer. The value of $s(t)$ at the first data point ($m = 0$) thus equals the presumed value at the $(N + 1)$st data point ($m = N$)). The Fourier transform can thus be construed as positioning the data points along a circular ring of temporal extent $N \Delta t$. The FFT is therefore most precise if applied to signals that physically possess this periodicity, typified by the sine function above that is periodic over every 65 data points, assuming that this signal is a restricted sample of an infinite sine function. In contrast, for

```
signalR = sin( 10.5 * pi / 64 * [0 : 63] )
plot( signalR );
plot( imag ( fft ( signalR ) ) );
```

the FFT exhibits broad extrema, since the periodic extension of the signal contains large discontinuities at the computational window edges. To eliminate this discontinuity, albeit at the cost of slightly distorting low-frequency spectral components, the signal can be multiplied by a window function prior to Fourier transforming. An example of such a window function is the Hamming window function,

$$\tilde{s}_m = s_m \left[\frac{1}{2} \left(1 - \cos\left(\frac{2\pi m}{N} \right) \right) \right], \qquad m = 0, \quad 1, \quad \dots, \quad N - 1 \qquad (21.8)$$

which is explicitly periodic with period $N \Delta t$. Multiplying the signal by $\tilde{s}$ before applying the FFT yields a frequency-spectrum distribution that more closely resembles that of a signal with this periodicity:

```
signalR = sin( 10.5 * pi / 64 * [0 : 63] ) * hamming( 64 )';
plot( imag ( fft ( signalR ) ) );
```

High-frequency, noisy components are eliminated from a signal by Fourier transforming, applying a filter in the Fourier domain and then inverse transforming. To illustrate, noise can be incorporated into a sine-function signal by adding a uniformly distributed random number in the interval $[-0.6, 0.6]$ to each signal value:

```
signalR = sin( 10 * pi / 64 * [0 : 63] ) + ...
    1.2 * ( rand(1, 64) - 0.5 );
plot( signalR );
```

This results in a fluctuating but, on average, frequency-independent contribution to the FFT of the signal, as is evident from

```
fftSignalR = fft( signalR );
plot( abs ( fftSignalR ) );
```

To filter the noise, the negative and positive low-frequency components can be first translated to the center of the computational window through the **fftshift** operation and high frequencies suppressed through multiplication by the Hamming window function. The original ordering of the frequency spectrum is recovered by a second call to **fftshift** and the inverse FFT is employed to obtain the smoothed, signal:

```
fftShiftR = fftshift( fftSignalR );
fftFilterShiftR = fftShiftR .* hamming( 64 )';
fftResultR = fftshift( fftFilterShiftR );
plot( real( ifft( fftResultR ) ) );
```

Chapter 22
Differential equations

A differential equation can be viewed as a rule that relates the value of a single variable at a point in space or time to the immediately preceding values of the variable. While analytic procedures can be employed to determine the global solution from the local relation, discretizing the space or time variable transforms the differential equation into a difference equation. The particle trajectory can then be obtained by repeatedly advancing the position and velocity numerically over small time intervals. From the values of the variable at infinitesimally displaced initial points (the *initial conditions*) the rule embodied in the difference equation can be iterated to determine the global behavior of the variable. For example, from the second-order differential equation describing the local relationship between the force and the acceleration on a point particle, the particle's location and velocity are determined for all future times once the values of these quantities at an initial time, or equivalently the particle locations at two infinitesimally separated initial times, are specified.

22.1 Euler's method

Euler's method recasts a linear Nth-order differential equation as a set of N coupled first-order differential equations followed by application of the forward finite-difference approximation to each equation in the set. If the initial values of the variables are specified, a time-stepping procedure yields their values at future times. For example, Newton's differential equation for a single massive particle with mass m attached to a spring with force constant k,

$$a = \frac{d^2x}{dt^2} = -\frac{k}{m}x \tag{22.1}$$

is written as a coupled set of first-order differential equations by introducing the velocity:

$$v = \frac{dx}{dt}$$

$$a = \frac{dv}{dt} = -\frac{k}{m}x \tag{22.2}$$

Analogously, the third-order differential equation

$$\frac{d^3x}{dt^3} = -kv \tag{22.3}$$

becomes

$$v = \frac{dx}{dt}$$
$$a = \frac{dv}{dt} \tag{22.4}$$
$$\frac{da}{dt} = -kv$$

Replacing the first-order derivatives by the forward difference approximation $D^+_{\Delta t}$ in Eq. (22.2) yields

$$\frac{x(t + \Delta t) - x(t)}{\Delta t} = v(t) + O(\Delta t)$$
$$\frac{v(t + \Delta t) - v(t)}{\Delta t} = -\frac{k}{m}x(t) + O(\Delta t) \tag{22.5}$$

or, equivalently,

$$x(t + \Delta t) = x(t) + \Delta t\, v(t) + O(\Delta t)^2$$
$$v(t + \Delta t) = v(t) - k\,\Delta t\, x(t)/m + O(\Delta t)^2 \tag{22.6}$$

where the product of the $O(\Delta t)$ error terms in Eq. (22.5) with Δt reduces the error order to $O(\Delta t)^2$. C | | code that retains the above variable names for $k = m = 1$, $\Delta t = 0.06$ and initial conditions $x(0) = 0$, $v(0) = 1$ is given for the DISLIN plotting package by

```
const NUMBEROFTIMESTEPS = 100;

main( ) {
      double x[NUMBEROFTIMESTEPS], v[NUMBEROFTIMESTEPS], k=1,
          m=1, dt=0.06;
      x[0] = 0;
      v[0] = 1;// second-order equation → two boundary conditions
      for ( int loop = 1; loop < NUMBEROFTIMESTEPS; loop++ ) {
          x[loop] = x[loop - 1] + dt * v[loop - 1];
          v[loop] = v[loop - 1] - k * dt * x[loop - 1] / m;
      }
      qplot( x, v, NUMBEROFTIMESTEPS )
}
```

The above program can be immediately adapted to a ball in the presence of a drag force proportional to velocity,

$$\frac{d^2\vec{r}}{dt^2} = -g\hat{e}_z - \alpha\vec{v} \tag{22.7}$$

In Octave, for a ball launched from coordinate origin with x and z components of velocity given by 10 and 100 m/s

```
numberOfTimeSteps = 100;
deltaTime = 0.2;
gravitationalConstant = -9.8;
dragConstant = 6.0E-2;

Ball.positionRC = zeros( 2, numberOfTimeSteps );
Ball.velocityRC = zeros( 2, numberOfTimeSteps );
% initial x and z velocities
Ball.velocityRC(:, 1) = [ 10; 100 ];

for loop = 2 : numberOfTimeSteps
        force = [ 0 ; gravitationalConstant ] - ...
            dragConstant * Ball.velocityRC(:, loop - 1);
        Ball.positionRC(:, loop) = Ball.positionRC(:, loop - 1) + ...
            deltaTime * Ball.velocityRC(:, loop - 1);
        Ball.velocityRC(:, loop) = Ball.velocityRC(:, loop - 1) + ...
            deltaTime * force;
end

plot( Ball.positionRC(1, :), Ball.positionRC(2, :));
```

If the Euler method is applied to the equation $dx/dt = f(t)$, each evaluation of $x(t + \Delta t)$ adds an increment given by $f(t)\Delta t$ to the previous value of $x(t)$, corresponding to the rectangular integration rule. Thus, the Euler procedure effectively generalizes rectangular integration to systems of first-order differential equations.

Newton's equations can also be programmed directly in second-derivative form by replacing the continuous second-derivative operator by its centered finite-difference operator $(D_{\Delta t}^0)^2$ approximation:

$$\left(D_{\Delta t}^0\right)^2 f(t) \equiv \frac{1}{\Delta t}\left(D_{\Delta t}^0\big|_{t+\Delta t/2} - D_{\Delta t}^0\big|_{t-\Delta t/2}\right) f(t)$$

$$= \frac{1}{(\Delta t)^2}([f(t + \Delta t) - f(t)] - [f(t) - f(t - \Delta t)])$$

$$= \frac{1}{(\Delta t)^2}(f(t + \Delta t) - 2f(t) + f(t - \Delta t)) \tag{22.8}$$

Inserting this expression directly into Eq. (22.1) yields a solution algorithm for $x(t + \Delta t)$ in terms of the initial conditions $x(t)$ and $x(t - \Delta t)$.

22.2 Error analysis

While $x(t + \Delta t)$ in Eq. (22.6) is accurate to $O(\Delta t)^2$, the particle position is normally required after a fixed propagation time T, corresponding to $T/\Delta t$ evolution steps. If the error per step is $O(\Delta t)^2$, the total error is therefore, as in numerical integration, $O(\Delta t)^2(T/\Delta t)$ or $O(\Delta t)$. However, the Euler method yields artificial divergences that increase with Δt in non-dissipative problems, as illustrated in

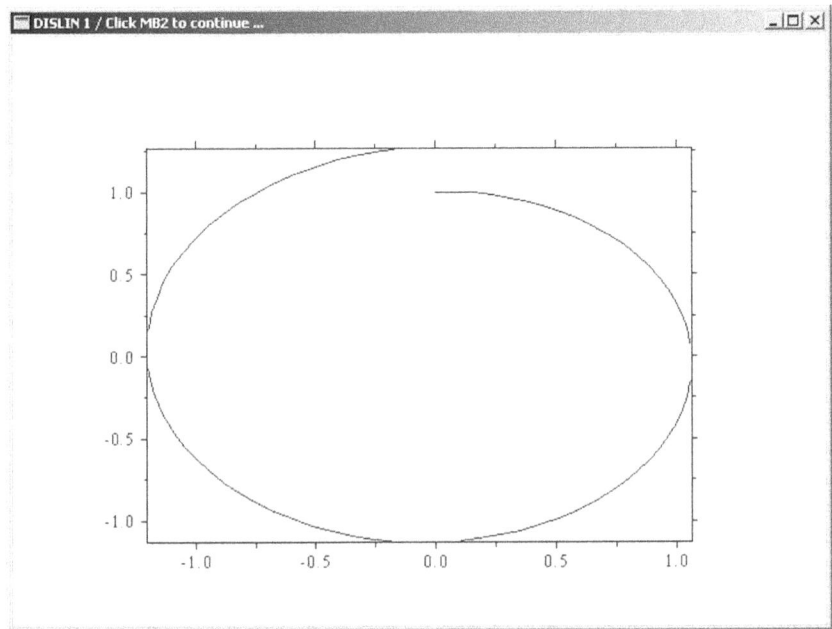

Figure 22.1

the phase-space plot, Figure 22.1, of velocity v against position x for 80 time steps with $\Delta t = 2\pi/80$.

The monotonic increase in spring energy occurs because, since the particle initially propagates from $x = 0$ to $x = 1$, the position at which the restoring force is evaluated, $x(t_i)$, is closer to the origin than the average displacement, $\approx x(t_i + \Delta t/2)$ of the spring during the propagation interval. Similarly, the value of the velocity variable in the first line of Eq. (22.6) that controls the magnitude of the change in the displacement over a step is larger than its properly averaged effective value over the time interval. Accordingly, the restoring force is weakened while the velocity is overestimated, so the displacement after the first quarter-cycle exceeds its correct value. In contrast, as the particle returns back to zero displacement, the negative restoring force is instead overestimated and the magnitude of the negative velocity term underestimated. Therefore over this quarter-cycle a stronger restoring force operates over a longer time, amplifying the magnitude of the negative velocity after a half-period, as again evidenced in Figure 22.1. By extension, the numerical error supplies a fictitious numerical driving force in resonance with the motion, which, in the absence of a physical dissipation mechanism, induces a steady growth of the particle energy.

Consistent with the above analysis, the unphysical energy divergence can be eliminated by balancing the error in the force in Eq. (22.6) against that of the velocity term. The resulting *Euler–Cromer* procedure is obtained by replacing

$v(t)$ in the first line by $v(t + \Delta t)$:

$$x(t + \Delta t) = x(t) + \Delta t \, v(t + \Delta t) + O(\Delta t)^2$$

$$v(t + \Delta t) = v(t) - k \, \Delta t \, x(t)/m + O(\Delta t)^2$$

(22.9)

The resulting cancellation in the two error contributions can be shown through algebraic manipulations to restore energy conservation

22.3   The Runge–Kutta procedure

The *Runge–Kutta* methods provide another set of time-stepping procedures for equations of the form

$$\frac{d\vec{x}}{dt} = \vec{f}(\vec{x}(t), t)$$

(22.10)

where for an n-dimensional system $\vec{f}$ is a vector of n position and n velocity coordinates. Specializing to a single dimension and a scalar function f, the procedure is obtained from the observation that

$$\frac{x(t + \Delta t) - x(t)}{\Delta t} = \frac{dx}{dt} + \frac{\Delta t}{2!} \frac{d}{dt}\left(\frac{dx}{dt}\right) + O(\Delta t)^2$$

$$= f(x(t), t) + \frac{\Delta t}{2} \frac{d}{dt}(f(x(t), t)) + O(\Delta t)^2$$

$$= f(x(t), t) + \frac{\Delta t}{2}\left(\frac{d}{dt}(f(x(t), t)) + \underbrace{\frac{dx(t)}{dt}}_{f(x(t),t)} \frac{d}{dx}(f(x(t), t)) \right)$$

$$+ O(\Delta t)^2$$

(22.11)

However, the last version can be written

$$(c_2 + c_3)f + c_1 c_3 \, \Delta t(f f_x + f_t) + O(\Delta t)^2$$

(22.12)

if $c_2 + c_3 = 1$, $c_1 c_3 = 1/2$. Rearranging the resulting expression gives

$$= c_2 f + c_3[f + c_1 \, \Delta t \, f f_x + c_1 \, \Delta t \, f_t + O(\Delta t)^2]$$

$$= c_2 f + c_3[f(x(t) + c_1 \, \Delta t \, f, t) + c_1 \, \Delta t \, f_t(x(t) + c_1 \, \Delta t \, f, t) + \tilde{O}(\Delta t)^2] \quad (22.13)$$

$$= c_2 f(x(t), t) + c_3 f(x(t) + c_1 \, \Delta t \, f(x(t), t), t + c_1 \, \Delta t) + \tilde{O}(\Delta t)^2$$

For e.g. $c_2 = c_3 = 1/2$, $c_1 = 1$ and the right-hand side of Eq. (22.11) is replaced by the average of the function evaluated at the left- and estimated right-hand endpoints of the interval.

Carrying this analysis further yields the fourth-order accurate procedure

$$\vec{x}(t + \Delta t) = \vec{x}(t) + \frac{1}{6} \, \Delta t(\vec{f}(\vec{x}(t), t) + 2\vec{F}_2 + 2\vec{F}_3 + \vec{F}_4)$$

(22.14)

where

$$\vec{F}_2 = \vec{f}\left(\vec{x} + \frac{\Delta t}{2}\,\vec{f}(\vec{x}(t), t), t + \frac{\Delta t}{2}\right)$$

$$\vec{F}_3 = \vec{f}\left(\vec{x} + \frac{\Delta t}{2}\,\vec{F}_2, t + \frac{\Delta t}{2}\right)$$

$$\vec{F}_4 = \vec{f}\left(\vec{x} + \Delta t\,\vec{F}_3, t + \Delta t\right)$$

Runge–Kutta techniques are present in virtually every numerical program library.

Chapter 23
Monte Carlo methods

Monte Carlo procedures provide a standardized framework for solving numerous involved calculations in which algorithmic error is replaced by statistical error. These methods therefore prove particularly effective when applied to stochastic physical systems. Several illustrations of Monte Carlo techniques are presented below.

23.1 Monte Carlo integration

To integrate a general function over an complicated integration region in N dimensions with a Monte Carlo approach, the integration region is enclosed within a reference region of known area. Points are generated randomly inside the reference region and the ratio of the number of points that fall into the integration region to the total number of sample points is computed. Multiplying this ratio by the volume of the reference region leads to an estimate of the integral.

We illustrate the procedure for an arbitrary one-dimensional function, **f( )**, here set to a cosine function, over the interval $[0, \pi/2]$. The routine then generates random samples, (x_i, y_i) within a rectangular reference region defined by $[0, \pi/2]$ in x and by $[-2, 2]$ in y, which encompasses both the minimum and the maximum value of $f(x)$. The variable **numberOfPointsBelowCurve** is incremented by one for each random point that lies below $f(x_i)$. The ratio of the final value of this variable to the total number of sample points multiplied by the area of the reference region is then computed and the result added to the difference in the x-limits multiplied by the lower y-limit to yield the integral,

```
f = @cos;
numberOfPoints = 10000;
lowerLimit = 0;
upperLimit = pi / 2;
% Must be smaller than the minimum value of f in the interval
yLower = -2;
% Must be greater than the maximum value of f in the interval
yUpper = +2;
regionArea = abs( yUpper - yLower ) * ( upperLimit - lowerLimit );
```

```
% Place numberOfPoints points randomly within the reference region
xValuesR = rand( 1, numberOfPoints ) * ...
   ( upperLimit - lowerLimit ) + lowerLimit;
yValuesR = rand( 1, numberOfPoints ) * ( yUpper - yLower ) + yLower;
% Count the number of points that fall below the integrand function
numberOfPointsBelowCurve = sum( yValuesR < f( xValuesR ) );
integral = numberOfPointsBelowCurve * ...
   regionArea / numberOfPoints +  yLower * ( upperLimit - lowerLimit )
```

23.2 Monte Carlo evaluation of distribution functions

As noted above, Monte Carlo techniques are particularly suited to the evaluation of statistical quantities such as probability distribution functions. Typically, these relate to a physical system composed of N subsystems, each of which is characterized by a randomly varying *local* parameter, s_i. This local parameter could be a spin that can only possess certain discrete values, a single flip of a coin that is sampled multiple times, or a continuous variable such as the impedance of an individual component of a transmission system. Since the local parameters are generally not deterministic, but instead fluctuate among subsystems or realizations as a result, for example, of thermal effects, manufacturing uncertainties or component aging, they are termed *stochastic variables*. For each set of local parameter values one or more *global* variables, such as a pulse propagation time, total series resistance or sample magnetization, are typically measured or numerically modeled. The problem is then to predict the probability that the global variables possess certain values when averaged over all realizations. Practically significant properties such as the average magnetization, mean time to failure or bit-error rate can then be determined.

A stochastic system can be modeled by assigning sets of random values to the local variables in a manner consistent with their statistics. For each set or *realization* the M global variables, $E^{(1)}, E^{(2)}, \ldots, E^{(M)}$ of interest are calculated. This result is then placed in the corresponding bin B of an M-dimensional histogram, which records the total number of realizations for which the $E^{(i)}$ are located within a certain restricted region in the global variable space. The distribution of events in the histogram for a large number of realizations corresponds to the desired probability distribution function. Mathematically, if $I_B(\vec{s}_i)$ is one within a histogram bin B and zero outside, then the probability distribution function after N_R realizations is given for this bin by

$$p_B(\vec{E}) = p(\{E^{(1)}, E^{(2)}, \ldots, E^{(M)}\} \in B) = \frac{1}{N_R} \sum_{i=1}^{N_R} I_B(\vec{s}_i) \qquad (23.1)$$

The above considerations are illustrated by a one-dimensional unbiased random walk. The underlying system variables s_i are assigned either 0 and 1 with equal probabilities corresponding to a displacement to the left or right by a unit distance. The global quantity of interest, E, is the total displacement after N steps. Averaging E over many realizations of the system variables yields a discrete distribution function of the probability that the walk terminates at a given displacement from the origin, as in

```
clear all
numberOfSteps = 40;
numberOfRealizations = 50000;
% For 2 steps, the possible outcomes are -2, 0 and 2,
% necessitating the +1 below
histogramR = zeros( 1, numberOfSteps + 1 );
for loop = 1 : numberOfRealizations
    % Simulate numberOfSteps random steps
    % 0 = step to left, 1 = step to right,
    % histogramIndex = number of right steps + 1
    histogramIndex = sum( round( rand( 1, numberOfSteps ) ) ) + 1;
    histogramR(histogramIndex) = histogramR(histogramIndex) + 1;
end
% Normalize distribution to unit sum
histogramR = histogramR / sum( histogramR );
% xScaleR = number of steps to right - number of steps to left
xScaleR = 2 * ( 0 : numberOfSteps ) - numberOfSteps;
semilogy( xScaleR, histogramR, "o", 'markersize', 3 );
```

Since the probability distribution function is approximately Gaussian (e.g. of the form $a \exp(-x^2/(2a^2))$), graphing the logarithm of the distribution function displays an inverted parabola. Slight distortions of the parabola in highly sampled regions indicate residual correlations within **rand( )** and are eliminated by using improved random-number algorithms.

23.3 Importance sampling

Estimating the probability of infrequent but physically important events such as, for example, those associated with system malfunctions requires many Monte Carlo realizations. If, however, the ranges of the local parameter values that yield these low-probability events are known from physical or mathematical considerations, the realizations can often be weighted to generate relevant sets of local variables more frequently than their occurrence in random samples. The resulting biased histogram, $I_B(\vec{s})$, for the global variables, however, overestimates the probability of such events and must be multiplied by a likelihood ratio

$$L(\vec{s}) = p(\vec{s})/p_B(\vec{s}) \tag{23.2}$$

that is often determined analytically as a function of the local parameters and expresses the quotient of the probabilities of the unbiased, $p(\vec{s})$, and the biased

distributions. The physical probability density distribution is then obtained
from

$$p(\vec{E}) = \frac{1}{N_R} \sum_{i=1}^{N_R} I_B(\vec{s}_i) L(\vec{s}_i) \tag{23.3}$$

For the above one-dimensional random walk, the local steps can be biased so
that the probability of moving to the right is 0.6 while that of moving to the left
is 0.4. (Note that, if the random walker does in reality favor moves to the right,
the resulting distribution would be the unbiased solution to the problem.) The
resulting histogram is then oversampled for rightward displacements. However,
the unbiased distribution can be generated by multiplying this distribution by 0.4
for each rightward step and 0.6 for each leftward step, as in the program below:

```
clear all
numberOfSteps = 40;
numberOfRealizations = 50000;
% For 2 steps, the possible outcomes are -2, 0 and 2,
% necessitating the +1 below
histogramR = zeros( 1, numberOfSteps + 1 );
for loop = 1 : numberOfRealizations
        % Simulate numberOfSteps random steps:
        % +1 = right step, 0 = left step
        % Events biased so that P(1) = 0.6
        histogramIndex = sum( round( rand( 1, numberOfSteps ) ) ) + 1;
        histogramR(histogramIndex) = histogramR(histogramIndex) + 1;
end
% Unbias result by multiplying by 0.6 for each left move
% and 0.4 for each right move
for loop = 1 : numberOfSteps;
        histogramR(loop) = histogramR(loop) * ...
                0.4^loop * 0.6^( numberOfSteps - loop );
end
% Normalize distribution to unit sum
histogramR = histogramR / sum( histogramR );
xScaleR = 2 * ( 0 : numberOfSteps ) - numberOfSteps;
semilogy( xScaleR, histogramR, 'x', 'markersize', 3 );
```

The results for the biased (circles) and unbiased (crosses) procedures are presented in Figure 23.1.

23.4 The Metropolis algorithm

While importance sampling can be applied only to systems for which the relationship between the local variables and global variable values has been carefully
analyzed, *statistical* biasing methods that maximize the entropy of the calculation
(e.g. minimize the amount of specialized information introduced or required by
the numerical method) can be universally applied. These procedures are often

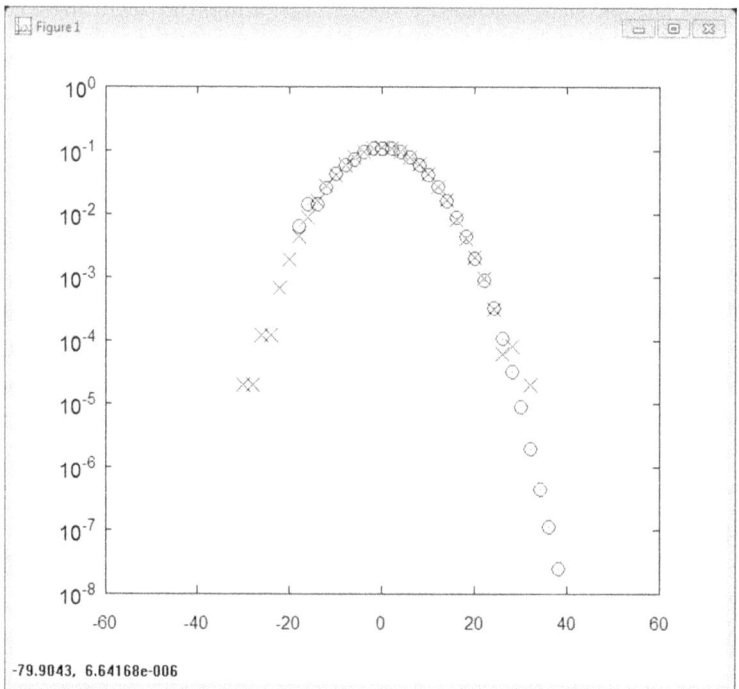

Figure 23.1

based on *Markov chains*, for which each new system realization slightly modifies the previous realization. An appropriate rule governs the acceptance or rejection of each such transition. If a transition is rejected, the histogram is incremented according to the previous result.

The first such, *Metropolis*, technique increases the probability of a global system variable, E, by a factor $e^{\beta E}$ (or, with trivial modifications, $e^{-\beta E}$). The degree of bias toward large E can therefore be adjusted by varying β. In the method, a Markov chain is first generated by assigning random values to the local system variables. One or more of these variables is then perturbed in such a manner that the global variables are changed by a small amount. The rule then accepts all transitions that increase the value of E, while permitting transitions that lower E by an amount $\Delta E < 0$ with a probability $e^{\beta \Delta E}$. If a transition is rejected, the previous sample is counted again. Since there always exists a finite probability for a transition to smaller E, states in the Markov chain can escape from local maxima, although only after an average number of transitions that increases exponentially with the height of the maximum relative to its surroundings.

To understand the origin of the transition rule, consider a system with two equally probable global states with different E. The probability of the Markov chain transitioning from the state with smaller E to the state with $E + \Delta E$ is then unity, while the probability of a transition in the reverse direction is $e^{-\beta \Delta E}$. However, on average, the number of transitions in one direction must be equal to the number of transitions in the other direction, otherwise the number of visits to the state with more incoming transitions would increase until equality is established. Accordingly, the average number of times the upper state is visited will exceed the corresponding number for the lower state by the factor $e^{\beta \Delta E}$. Since this argument extends to any distribution of global states, the Metropolis algorithm oversamples a state with global variable E with a probability proportional to $e^{\beta E}$. Therefore the physical probability distribution function (in statistical mechanics the density-of-states function) is obtained by multiplying the resulting probability distribution by the likelihood function $L(E) = e^{-\beta E}$.

The following Metropolis random-walk program implements successive realizations in a Markov chain formed from a random walk with **numberOfSteps** steps by changing the direction of a single, randomly chosen step. The global variable E is set to the absolute value of the distance of the final position of the random walk from the origin. If a transition is rejected, the histogram bin associated with the previous transition is incremented by unity, which is termed a self-transition. The final histogram is multiplied by the likelihood function to generate the unbiased probability distribution:

```
clear all
numberOfRealizations = 100000;
numberOfSteps = 40;
myBeta = 1.0;
histogramR = zeros( 1, numberOfSteps + 1 );
% Simulate numberOfSteps coin flips (or random steps)
stepSequenceR = round( rand( 1, numberOfSteps ) );
histogramIndex = sum( stepSequenceR ) + 1;
centralPoint = numberOfSteps / 2;
for loop = 1 : numberOfRealizations;
    flipPosition = fix( rand * numberOfSteps ) + 1;
    stepSequenceR(flipPosition) = 1 - stepSequenceR(flipPosition);
    histogramIndexNew = sum( stepSequenceR ) + 1;
    if rand < exp( myBeta *( ...
            abs( histogramIndexNew - centralPoint ) - ...
            abs( histogramIndex - centralPoint ) ) )
            histogramIndex = histogramIndexNew;
    else
            stepSequenceR(flipPosition) = ...
                1 - stepSequenceR(flipPosition);
    end
    histogramR(histogramIndex) = histogramR(histogramIndex) + 1;
end
histogramR = histogramR .* exp( - myBeta * ...
    abs( [1 : numberOfSteps + 1] - centralPoint ) );
xScaleR = 2 * ( 0 : numberOfSteps ) - numberOfSteps;
semilogy( xScaleR, histogramR / sum( histogramR ) );
```

Executing the above program for values of **myBeta** from e.g. 0.2 to 1 after issuing the command **hold on** demonstrates the influence of the biasing function. For large **myBeta**, the sample space becomes overbiased toward statistically unlikely events, undersampling the high-probability region of the probability distribution function. However, the probability distribution function can be evaluated for several values of **myBeta** and each result employed in its region of greatest accuracy.

23.5 Multicanonical methods

As demonstrated in the previous section, while Metropolis and importance-sampling methods bias events toward certain global variable regions, statistical accuracy is lowered elsewhere. An improved choice for the bias function instead populates all histogram bins with approximately equal numbers of events. This is achieved if successive realizations of a Markov chain execute an unbiased random walk in the space of the global variables, E. Successive realizations then can escape even from pronounced extrema. However, in order for the biased probability distribution function not to depend on E, the likelihood ratio, Eq. (23.2), must be proportional to the desired probability distribution function. While this function is the result of the procedure and is therefore by definition not initially known, it can be determined iteratively such that an unbiased Markov chain calculation generates an initial Monte Carlo distribution function that is employed to bias samples of the subsequent iteration toward regions of small probability.

To implement the procedure, two histograms, h and H, with elements that are initialized to unity, are introduced. The first stores the iteratively updated probability distribution function and the second the intermediate statistics for the current iteration that are required in order to update h. The distribution function is initially taken as a constant function, for which all elements of h_0 are unity. Next, a Markov-chain simulation is performed, with the rule that transitions that increase the current estimate of h are accepted with a probability $h_0(\vec{E}_n)/h_0(\vec{E}_{n+1})$ and those that decrease h with unit probability; in the initial step this yields a Monte Carlo distribution function. The histogram obtained from this simulation, is stored in H, which is designated H_0 in the initial step. Subsequently,

$$h_{i+1} = h_i H_i \qquad (23.4)$$

with $i = 0$ yields the first approximation, h_1, to the desired probability distribution function. H is then reinitialized to unity and the procedure is repeated with the new approximation for the distribution function, h_1, in place of h_0, and subsequently iterated.

Returning to the two-state system, suppose that the mth estimate, h_m, of the probability density function underestimates the ratio of the probabilities of the two states by a factor of 2. In the subsequent iteration, the transition rule suppresses transitions out of the underestimated state by this factor while the probability of an incoming transition remains unchanged. As a result, the underestimated state is sampled twice as many times as the overestimated state, resulting in an accurate estimate of the probability distribution function when H and h_m are multiplied. Additionally, as noted above, successive iterations sample increasingly lower-probability regions of the distribution function.

A multicanonical random-walk program is given below (while the distribution function is nearly independent of the method parameters over wide ranges, typically the size of the perturbations is adjusted so that approximately a quarter of the transitions are rejected):

```
clear all
numberOfRealizations = 100000;
numberOfSteps = 40;
numberOfOuterLoops = 4;
histogramR = ones( 1, numberOfSteps + 1 );
stepSequenceOldR = round ( rand( 1, numberOfSteps ) );
stepSequenceNewR = stepSequenceOldR;
histOld = 1;
histogramIndexOld = sum( stepSequenceOldR ) + 1;

for loopOuter = 1 : numberOfOuterLoops;
     histogramNewR = ones( 1, numberOfSteps + 1 );
     for loop = 1 : numberOfRealizations;
          stepSequenceNewR = stepSequenceOldR;
          % Change one local variable (step direction)
          flipPosition = ceil( numberOfSteps * rand );
          stepSequenceNewR(flipPosition) = ...
              1 - stepSequenceOldR(flipPosition);
          % Compute global variable (total distance traveled)
          histogramIndexNew = sum( stepSequenceNewR ) + 1;
          histNew = histogramR(histogramIndexNew);
          % Multicanonical acceptance rule
          if ( rand < histOld / histNew )
               stepSequenceOldR = stepSequenceNewR;
               histOld = histNew;
               histogramIndexOld = histogramIndexNew;
          end
     histogramNewR(histogramIndexOld) = ...
          histogramNewR(histogramIndexOld) + 1;
     end
     histogramR = histogramR .* histogramNewR;
end
xScaleR = 2 * ( 0 : numberOfSteps ) - numberOfSteps;
semilogy( xScaleR, histogramR / sum( histogramR ) );
```

23.6 Particle simulations

Molecular dynamics evolves a randomly generated set of energetic particles according to Newton's laws of motion. When two particles are sufficiently close, a scattering event is simulated by, for example, reorienting the particle motion along random outgoing angles with new velocities and energies. Monitoring the evolution of individual molecules or of molecules in selected regions yields the statistical medium properties.

As a simple one-dimensional illustration, 20 equal-mass point particles with random but successively ordered positions are generated within a region between 0 and 500 distance units terminated with reflecting boundaries. The particles are assigned random velocities between -1.0 and $+1.0$. The minimum time for a collision to occur, either of a particle with its neighbor to its immediate left or, for the first and last particle, with a boundary, is calculated. Every particle is then evolved over this minimum time. If the minimum is associated with a particle collision, the velocities of the colliding particles are interchanged; otherwise, the velocity of the particle at the boundary is reversed. This sequence of steps is repeated for 200 collisions and the trajectory of the central particle over this time is graphed:

```
clear all
numberOfParticles = 20;
numberOfCollisions = 200;
windowSize = 500;
collisionTime = 0;
% Assign random positions to the particles
% and order the positions in an array
Particle.positionR = rand( 1, numberOfParticles ) * windowSize;
% Sort particles from left to rightmost in the array
Particle.positionR( : ) = sort( Particle.positionR( : ) );
% Assign random velocities from -1.0 to +1.0 to each particle
Particle.velocityR = ( rand( 1, numberOfParticles ) - 0.5 ) * 2;
for loop = 1 : numberOfCollisions
  minimumTime = 1.e20;
  for loopParticle = 2 : numberOfParticles
% Check whether the particle to the left will collide
% with the current particle
    if Particle.velocityR(loopParticle - 1) - ...
       Particle.velocityR(loopParticle) > 0
% Compute the time until collision
      deltaTime = ( Particle.positionR(loopParticle) - ...
        Particle.positionR(loopParticle - 1) ) / ...
        ( Particle.velocityR(loopParticle - 1) - ...
        Particle.velocityR(loopParticle) );
% If this is smaller than previously recorded times,
% store the time and the particle number
        if deltaTime < minimumTime
        minimumTime = deltaTime;
        particleIndex = loopParticle;
% Set flag to indicate that this is not a wall collision
        isWall = 0;
        end;
      end
  end
% Compute similarly the time for the rightmost
% particle to collide with the right wall
  if Particle.velocityR(numberOfParticles) > 0
    deltaTime = ( windowSize - Particle.positionR(numberOfParticles) ...
      ) / Particle.velocityR(numberOfParticles);
```

```
    if deltaTime < minimumTime;
       minimumTime = deltaTime;
       particleIndex = numberOfParticles;
% Set the wall flag to indicate a wall collision
       isWall = 1;
    end;
 end
% Repeat this for the leftmost particle
 if Particle.velocityR(1) < 0
    deltaTime = -Particle.positionR(1) / Particle.velocityR(1);
    if deltaTime < minimumTime;
       minimumTime = deltaTime;
       particleIndex = 1;
       isWall = 1;
    end;
 end;
 for loopParticle = 1 : numberOfParticles
% Advance the particle positions over the time minimumTime
    Particle.positionR(loopParticle) = ...
      Particle.positionR(loopParticle) + ...
      Particle.velocityR(loopParticle) * minimumTime;
% If the colliding particle intersected a wall,
% change the sign of its velocity. Otherwise exchange the
% velocities of the two colliding particles
    if loopParticle == particleIndex
       if isWall
          Particle.velocityR(loopParticle) = ...
            -Particle.velocityR(loopParticle);
       else
          saveVelocity = Particle.velocityR(loopParticle);
          Particle.velocityR(loopParticle) = ...
            Particle.velocityR(loopParticle - 1);
          Particle.velocityR(loopParticle - 1) = saveVelocity;
       end
    end
 end
% Plot the time evolution of the central particle
    collisionTime = collisionTime + minimumTime;
    saveTimeR(loop) = collisionTime;
    savePositionR(loop) = Particle.positionR(numberOfParticles / 2);
end
plot( saveTimeR, savePositionR );
```

23.7 The Ising model

The evaluation of thermodynamic quantities is most simply illustrated by a one-dimensional *Ising model* in which successive spins in a one-dimensional system interact according to the Hamiltonian

$$E = -J \sum_{m=1}^{N_s} \sigma_m \sigma_{m+1} \tag{23.5}$$

Periodic boundaries apply to spins arranged around the circumference of a circle, such that $\sigma_{N_s+1} = \sigma_1$. For a spin-1/2 system, each σ_m equals $\pm 1/2$ and all possible configurations can be obtained by converting an integer into a binary number, the digits of which specify the state of the corresponding spin, and subsequently incrementing the integer. Each distinct energy, E_l, is recorded in an array together while an accompanying array stores the number of times, N_{E_l}, that the energy is observed. In the canonical ensemble (Boltzmann distribution), the probability of occupation of a state of energy E is given by $\exp(-\beta E)$ with $\beta = 1/(k_B T)$, where k_B is the Boltzmann constant and T is the system *temperature*, so that the average energy of the system is given by

$$\bar{E} = \frac{\displaystyle\sum_{l=1}^{N_E} E_l N_{E_l} e^{-\beta E_l}}{\displaystyle\sum_{l=1}^{N_E} N_{E_l} e^{-\beta E_l}} = \frac{\displaystyle\sum_{l=1}^{N_E} E_l N_{E_l} e^{-\beta E_l}}{Z} \tag{23.6}$$

where

$$Z = \sum_{l=1}^{N_E} N_{E_l} e^{-\beta E_l} \tag{23.7}$$

is termed the *partition function*. The *specific heat capacity* per spin, c_V, is then

$$c_V = \frac{1}{N_s}\frac{\partial \bar{E}}{\partial T} = \frac{1}{N_s}\frac{\partial \beta}{\partial T}\frac{\partial \bar{E}}{\partial \beta} = -\frac{\beta^2}{N_s}\frac{\partial \bar{E}}{\partial \beta} = -\frac{\beta^2}{N_s}\frac{\partial}{\partial \beta}\left(\frac{\displaystyle\sum_{l=1}^{N_E} E_l N_{E_l} e^{-\beta E_l}}{\displaystyle\sum_{l=1}^{N_E} N_{E_l} e^{-\beta E_l}}\right)$$

$$= \frac{\beta^2}{N_s}\left(\frac{\displaystyle\sum_{l=1}^{N_E} E_l^2 N_{E_l} e^{-\beta E_l}\sum_{l=1}^{N_E} N_{E_l} e^{-\beta E_l} - \left(\sum_{l=1}^{N_E} E_l N_{E_l} e^{-\beta E_l}\right)^2}{\left(\displaystyle\sum_{l=1}^{N_E} N_{E_l} e^{-\beta E_l}\right)^2}\right) = \frac{\beta^2}{N_s}(\overline{E^2} - (\bar{E})^2) \tag{23.8}$$

The above formula, in which N_s is the number of spins, is employed to generate the specific heat of a periodic one-dimensional Ising system with $J = 4$ in the program below:

```
clear all
Ising.numberOfEnergies = 1;
Ising.energy(1) = 0;
Ising.numberOfStates(1) = 0;
numberOfSpins = 8;
numberOfRealizations = 2^numberOfSpins;
for loop = 1 : numberOfRealizations
% Convert each number from 0 to 2^N-1 into binary
% form to represent a set of spins
```

```
        spins = dec2bin( loop - 1, numberOfSpins );
        energy = 0;
        for innerLoop = 1 : numberOfSpins
% Compute the Ising energy (interaction constant = -1)
                energy = energy - str2num( spins(innerLoop) ) * ...
                        str2num( spins( mod( innerLoop, ...
                        numberOfSpins ) + 1 ) );
        end
% Increase the number of recorded energies at this
% value by 1 if the value has been obtained earlier;
% otherwise store a new energy value in Ising
        energyFound = 0;
        for innerLoop = 1 : Ising.numberOfEnergies
                if Ising.energy(innerLoop) == energy;
                        Ising.numberOfStates(innerLoop) = ...
                                Ising.numberOfStates(innerLoop) + 1;
                        energyFound = 1;
                end
                if energyFound == 1; break; end;
        end
        if energyFound == 0
                Ising.numberOfEnergies = Ising.numberOfEnergies + 1;
                Ising.energy(Ising.numberOfEnergies) = energy;
                Ising.numberOfStates(Ising.numberOfEnergies) = 1;
        end
end
% Calculate statistical quantities for the
% specified value of beta = 1/kT
myBeta = 0.05;
partitionFunction = 0;
for loop = 1 : Ising.numberOfEnergies
        partitionFunction = partitionFunction + ...
                Ising.numberOfStates(loop) * ...
                exp( -myBeta * Ising.energy(loop) );
end
averageEnergy = 0;
averageSquaredEnergy = 0;
for loop = 1 : Ising.numberOfEnergies
        averageEnergy = averageEnergy + Ising.energy(loop) * ...
                Ising.numberOfStates(loop) * ...
                exp( -myBeta * Ising.energy(loop) );
        averageSquaredEnergy = averageSquaredEnergy + ...
                Ising.energy(loop)^2 * Ising.numberOfStates(loop) * ...
                exp( -myBeta * Ising.energy(loop) );
end
averageEnergy = averageEnergy / partitionFunction;
averageSquaredEnergy = averageSquaredEnergy / partitionFunction;
specificHeatPerSpin = myBeta^2 * ( averageSquaredEnergy - ...
        averageEnergy^2 ) / numberOfSpins;
```

Chapter 24
Partial differential equations

The properties of a *continuous* physical system, such as its local displacement from its equilibrium position, can in principle vary arbitrarily from one spatial point to an adjacent point. Accordingly, one or more degrees of freedom exist at every spatial coordinate value. However, in order for the medium to remain connected, the properties at a given location are influenced by those at neighboring locations through restoring forces. The system properties at a point thus evolve according to an ordinary differential equation that is, however, coupled to similarly evolving equations at neighboring points. Assuming that the form of these equations is identical throughout the medium, the entire system of coupled ordinary differential equations can be described by a single *partial differential equation*. Since the partial differential equation regenerates the underlying coupled ordinary differential equations when discretized, numerical methods for solving coupled ordinary differential equations generalize immediately to partial differential equations.

24.1 Scientific applications

As a concrete example, consider a non-uniformly heated one-dimensional metal bar with a temperature distribution $T(x, t)$ discretized on an equally spaced set of grid points, $x_i = x_L + (i - 1)\Delta x$, $i = 1, 2, \ldots, N$. The temperature, $T(x_i, t_{j+1})$, at x_i at a time $t_{j+1} = t_j + \Delta t$, for small Δt, increases from $T(x_i, t_j)$ by amounts proportional to $(T(x_{i+1}, t_j) - T(x_i, t_j))$ and $(T(x_{i-1}, t_j) - T(x_i, t_j))$ from the differing temperatures at x_{i+1} and x_{i-1}, respectively, i.e.

$$T(x_i, t_{j+1}) - T(x_i, t_j) = \bar{c}(T(x_{i+1}, t_j) - 2T(x_i, t_j) + T(x_{i-1}, t_j)) \qquad (24.1)$$

in which the constant $\bar{c}$ is dimensionless but proportional to Δt and $1/(\Delta x)^2$ (for example the change in T varies at first linearly with $\Delta t = t_{j+1} - t_j$). Dividing both sides by Δt and simultaneously multiplying and dividing the right-hand side by Δx yields

$$\frac{T(x_i, t_{j+1}) - T(x_i, t_j)}{\Delta t} = \bar{c}\frac{(\Delta x)^2}{\Delta t}\frac{T(x_{i+1}, t_j) - 2T(x_i, t_j) + T(x_{i-1}, t_j)}{(\Delta x)^2} \qquad (24.2)$$

which in continuous form corresponds to the parabolic diffusion equation

$$\frac{\partial T(x,t)}{\partial t} = D\frac{\partial^2 T(x,t)}{\partial x^2} \tag{24.3}$$

The thermal diffusion constant D possesses units of squared distance over time. Given the previous discussion of ordinary differential equations, a unique solution of the above equation requires initial conditions $T(x,t_0)$ at each point as well as *boundary conditions* that quantify the influence of the surrounding medium on the extremities of the material at every time.

If the average of the temperatures to the immediate right and left of a grid point is less than the temperature at the grid point, the curvature of the temperature field is negative and heat flows away from the grid point toward the boundaries with a *velocity* proportional to the local curvature of the distribution. For the temperature distribution $T(x, t = 0) = T_0 + A\sin(\pi x/L)$ with $0 < x < L$ and boundary temperatures $T(0,t) = T(L,t) = T_0$ for all $t > 0$ the negative curvature and hence the velocity of the temperature decay is proportional to $T(x,t) - T_0$, ensuring that the shape of the distribution is preserved with time, since

$$\frac{d}{dt}(\sin(\pi x/L)T(t)) = D\frac{d^2}{dx^2}(\sin(\pi x/L)T(t))$$
$$\frac{dT(t)}{dt} = -\frac{D\pi^2}{L^2}T(t) \tag{24.4}$$

so that the diffusion equation with zero boundary conditions is satisfied by

$$T(x,t) = e^{-D\pi^2 t/L^2}\sin(\pi x/L) \tag{24.5}$$

For a two-dimensional uniform temperature distribution $T(\vec{x},t)$, over small time intervals the temperature change at a given grid point is again proportional to the difference between its temperature and the average of the temperatures at the neighboring grid points. Assuming equal grid point spacing in the x- and y-directions, $\Delta x = \Delta y = \Delta\alpha$, Eq. (24.2) accordingly generalizes to

$$\frac{T(x_{i,k}, t_{j+1}) - T(x_{i,k}, t_j)}{\Delta t}$$
$$= \bar{c}\frac{(\Delta\alpha)^2}{\Delta t}\frac{T(x_{i+1,j}, t_j) + T(x_{i-1,j}, t_j) + T(x_{i,j+1}, t_j) + T(x_{x,j-1}, t_j) - 4T(x_i, t_j)}{(\Delta\alpha)^2} \tag{24.6}$$

yielding in the continuous limit the partial differential equation

$$\frac{\partial T(x,y,t)}{\partial t} = D\left(\frac{\partial^2 T(x,y,t)}{\partial x^2} + \frac{\partial^2 T(x,y,t)}{\partial y^2}\right) \tag{24.7}$$

Substituting *it* for t modifies the diffusion equation so that it describes wave motion instead of relaxation. The resulting *parabolic wave equation* is typified by the time-dependent Schrödinger equation

$$\frac{\partial\varphi}{\partial t} = i\frac{\hbar}{2m}\frac{\partial^2\varphi}{\partial x^2} \tag{24.8}$$

If $\varphi(x, 0) = A \sin(\pi x / L)$ for $0 < x < L$, the negative curvature initially yields a change in φ directed in the $-i$ direction. Since the velocity is proportional to the field magnitude, φ evolves with time into a imaginary *negative* sine function. At this point in time, its curvature is a *positive* imaginary quantity. The wavefunction thus experiences a negative real evolution with time, propagating along the negative horizontal direction in the complex plane. By extension, the field rotates in the complex plane so that both the imaginary part and the real part of the wavefunction oscillate with a 90° relative phase shift between them. Alternatively, since an eigenmode of the diffusion equation decays as $e^{-\alpha t}$, for $t \to it$, the field instead varies as $e^{-i\alpha t}$.

For hyperbolic equations of the form

$$\frac{\partial^2 S(x, t)}{\partial t^2} = D \frac{\partial^2 S(x, t)}{\partial x^2} \tag{24.9}$$

the *acceleration* rather than the velocity of the field is proportional to its curvature. For initial conditions $S(x, 0) = A \sin(\pi x / L)$ and zero boundary conditions, the acceleration of each spatial point possesses its maximum negative value at $t = 0$. A quarter of an oscillation period later $S(x, T/4) = 0$ and the field attains zero *acceleration* but a maximum negative velocity. The acceleration then reverses sign and becomes increasingly positive until the field distribution reaches the maximum negative displacement $A \sin(\pi x / L)$ consistent with conventional wave motion.

Finally *elliptic equations* are exemplified by the Poisson equation,

$$\frac{\partial^2 V(x, y, z)}{\partial x^2} + \frac{\partial^2 V(x, y, z)}{\partial y^2} + \frac{\partial^2 V(x, y, z)}{\partial z^2} = -\frac{1}{\varepsilon_0} \rho(x, y, z) \tag{24.10}$$

in which $-\rho/\varepsilon_0$, where ρ represents the charge density, is a source for the curvature of the electric potential V such that these two quantities are proportional at every point within the problem boundary. The electric field, given by the negative gradient of the potential, is visualized in Octave for a point charge at the origin with that $q/(4\pi\varepsilon_0) = 1$ as follows:

```
clear all
numberOfPoints = 6;                        % must be even
halfWidth = 5;                             % halfwidth of grid
xPositionR = linspace( -halfWidth, halfWidth, numberOfPoints );
yPositionR = xPositionR;
field.xValueRC = zeros( numberOfPoints, numberOfPoints );
field.yValueRC = zeros( numberOfPoints, numberOfPoints );
for outerLoop = 1 : numberOfPoints
    for innerLoop = 1 : numberOfPoints
        radius = sqrt( xPositionR(outerLoop)^2 + ...
            yPositionR(innerLoop)^2 );
        unitX = xPositionR(outerLoop) / radius;
        unitY = yPositionR(innerLoop) / radius;
        field.xValueRC(innerLoop, outerLoop) = unitX / radius^2;
```

```
        field.yValueRC(innerLoop, outerLoop) = unitY / radius^2;
    end
end

quiver( xPositionR, yPositionR, field.xValueRC, field.yValueRC );
```

Discretizing Poisson's equation in three dimensions yields

$$\frac{V(r_{i+1,j,k}) + V(r_{i-1,j,k}) + V(r_{i,j+1,k}) + V(r_{i,j-1,k}) + V(r_{i,j,k+1}) + V(r_{i,j,k-1}) - 6V(x_i)}{(\Delta\alpha)^2}$$

$$= -\frac{\rho(x_{i,j,k})}{\varepsilon_0} \tag{24.11}$$

For a unit point charge, $\rho(x, y, z) = \delta(\vec{r})$, the continuous solution for the potential is $V(\vec{r}) = 1/(4\pi\varepsilon_0 r)$. That the left-hand side of the above equation evaluates to zero except at the origin as $\Delta\alpha \to 0$ can be established with the discrete formalism. Since the charge distribution and therefore the potential are spherically symmetric, it suffices to evaluate the left-hand side of Eq. (24.11) for a point on the x-axis. This yields (noting that displacements in the y- and z-directions yield identical contributions by symmetry and recalling that $(1+\varepsilon)^\gamma \approx 1 + \gamma\varepsilon + \gamma(\gamma-1)\varepsilon/2! + \cdots$),

$$\frac{1}{(\Delta\alpha)^2} \left(\frac{1}{x+\Delta\alpha} + \frac{1}{x-\Delta\alpha} + \frac{4}{\sqrt{x^2+(\Delta\alpha)^2}} - \frac{6}{x} \right)$$

$$\approx \frac{1}{(\Delta\alpha)^2} \left(\frac{1}{x}\left(1 - \frac{\Delta\alpha}{x} + \left(\frac{\Delta\alpha}{x}\right)^2 - \left(\frac{\Delta\alpha}{x}\right)^3\right) \right.$$

$$+ \frac{1}{x}\left(1 + \frac{\Delta\alpha}{x} + \left(\frac{\Delta\alpha}{x}\right)^2 + \left(\frac{\Delta\alpha}{x}\right)^3\right)$$

$$\left. + \frac{4}{x}\left(1 - \frac{1}{2}\left(\frac{\Delta\alpha}{x}\right)^2\right) + O(\Delta\alpha)^4 - \frac{6}{x} \right) \tag{24.12}$$

which approaches zero as $(\Delta\alpha)^4$ for $\Delta\alpha \to 0$. The function $1/r$ accordingly possesses a positive curvature in the radial direction but a negative curvature in both of the orthogonal directions, such that the sum of these curvatures vanishes to second order.

In general a partial differential equation of the form

$$A\frac{\partial^2\xi}{\partial x^2} + B\frac{\partial^2\xi}{\partial x\,\partial y} + C\frac{\partial^2\xi}{\partial y^2} + D\frac{\partial\xi}{\partial x} + E\frac{\partial\xi}{\partial y} + F\xi + G = 0 \tag{24.13}$$

is termed elliptic, parabolic or hyperbolic depending on whether $B^2 - 4AC$ is $<$, $=$ or $>$ zero, respectively.

To obtain a unique solution to a partial differential equation, *initial conditions* must be specified at each point in space, while *boundary conditions* that quantify the interaction of the extremities of the field with its environment must be given at each boundary point for all times. *Dirichlet boundary conditions* specify the field value on the boundaries, while *Neumann boundary conditions* instead designate the flux (normal derivative) of the field out of the boundaries. *Mixed boundary conditions* impose Dirichlet conditions over part of the boundary and Neumann conditions over the remaining regions. If the material or the numerical implementation possesses a ring topology typified by a thin rod bent into a circle, the field obeys *periodic boundary conditions*, for example, $T(0, t) = T(L, t)$ and $dT(0, t)/dx = dT(L, t)/dx$ for temperature. The total number of boundary conditions is identical in all cases, since these must be specified at each boundary point in any number of dimensions.

24.2 Direct solution methods

Since a discretized partial differential equation yields a system of coupled ordinary differential equations, an initial field can be evolved in time through the Euler direct solution method that applies a forward difference approximation to the time derivative. For the diffusion equation

$$T(x_i, t_{j+1}) = T(x_i, t_j) + \frac{D \, \Delta t}{(\Delta x)^2} \left\{ T(x_{i-1}, t_j) - 2T(x_i, t_j) + T(x_{i+1}, t_j) \right\} \quad (24.14)$$

The above equation with an initial field $T(x, 0) = \sin(\pi x/L)$ and zero-temperature boundaries $T(x_0, t) = T(x_{N-1}, t) = 0$ can be implemented by advancing the field only over the points $i = 1, 2, \ldots N - 2$, while maintaining a constant (here zero) temperature at $i = 0, N - 1$. *Note that the temperature field at the current step must be saved when computing the updated field, otherwise $T(x_i, t_j)$ is overwritten before $T(x_{i+1}, t_{j+1})$ is computed. Failure to save fields in this manner is a common scientific programming error.* The program is

```
const int NUMBEROFPOINTS = 100;
main( ) {
    double deltaTime = 0.01;
    int numberOfTimeSteps = 5000;
    double deltaX = 0.1;
    double diffusionConstant = 0.5;
    double coefficient = diffusionConstant *
        deltaTime / ( deltaX * deltaX );
    float position[NUMBEROFPOINTS];
    float tempLast[NUMBEROFPOINTS], temp[NUMBEROFPOINTS ] = { 0 };
    for ( int loop = 0; loop < NUMBEROFPOINTS; loop++ ) {
        position[ loop ] = deltaX * loop;
        temp[loop] = sin( loop * M_PI /
            (NUMBEROFPOINTS - 1 ) );
    }
    for ( int outerLoop = 0; outerLoop <
        numberOfTimeSteps; outerLoop++ ) {
        if ( outerLoop % 500 == 0 ) qplot( position, temp,
            NUMBEROFPOINTS );
        for ( int loop = 1; loop < NUMBEROFPOINTS - 1; loop++ )
```

```
        tempLast[loop] = temp[loop];
  for ( int loop = 1; loop < NUMBEROFPOINTS - 1; loop++ )
      temp[loop] = coefficient * ( tempLast[loop - 1] - 2 *
          tempLast[loop] + tempLast[loop + 1] ) + temp[loop];
  }
}
```

The output of the program describes an undistorted but decaying sine function.

24.3 Hyperbolic differential equations and electromagnetics

Hyperbolic differential equations are typified by the wave equation, which in one dimension is

$$\frac{\partial^2 \chi}{\partial t^2} = v^2 \frac{\partial^2 \chi}{\partial x^2} \tag{24.15}$$

Just as the equation of motion for a spring, $\partial^2 x/\partial t^2 = -\omega^2 x$, can be expressed as two coupled one-dimensional equations

$$\frac{d}{dt}\begin{pmatrix} x \\ \tilde{v} \end{pmatrix} = \begin{pmatrix} 0 & \omega \\ -\omega & 0 \end{pmatrix}\begin{pmatrix} x \\ \tilde{v} \end{pmatrix} \quad <!--\tilde{v} = \sqrt{m/k}\, v = v/\omega -->\tag{24.16}$$

in which $\frac{\tilde{v}=v}{\omega} = v/\omega$ represents a modified velocity with units of distance and $\omega = \sqrt{k/m}$ is the ratio between the maximum velocity and the maximum displacement, the wave equation can be recast into the form

$$\frac{\partial}{\partial t}\begin{pmatrix} \chi \\ \xi \end{pmatrix} = \begin{pmatrix} 0 & v \\ v & 0 \end{pmatrix}\frac{\partial}{\partial x}\begin{pmatrix} \chi \\ \xi \end{pmatrix} \tag{24.17}$$

where, however, χ and ξ often represent fundamentally different physical quantities. For example, Maxwell's equations for the electric and magnetic fields in free space in the absence of charge and current sources for an x-polarized electric field that is uniform in the x–y plane and propagating in the z-direction so that $\vec{E} = E_x(z)\hat{e}_x$ become

$$\frac{\partial \vec{H}}{\partial t} = \frac{1}{\mu_0}\vec{\nabla} \times \vec{E} = \frac{1}{\mu_0}\begin{vmatrix} \hat{e}_x & \hat{e}_y & \hat{e}_z \\ \frac{\partial}{\partial x} & \frac{\partial}{\partial y} & \frac{\partial}{\partial z} \\ E_x(z) & 0 & 0 \end{vmatrix} = \frac{1}{\mu_0}\hat{e}_y\frac{\partial E_x(z)}{\partial z}$$

$$\tag{24.18}$$

$$\frac{\partial \vec{E}}{\partial t} = -\frac{1}{\varepsilon_0}\vec{\nabla} \times \vec{H} = -\frac{1}{\varepsilon_0}\begin{vmatrix} \hat{e}_x & \hat{e}_y & \hat{e}_z \\ \frac{\partial}{\partial x} & \frac{\partial}{\partial y} & \frac{\partial}{\partial z} \\ 0 & H_y(z) & 0 \end{vmatrix} = \frac{1}{\varepsilon_0}\hat{e}_x\frac{\partial H_y(z)}{\partial z}$$

Recalling that in vacuum the light velocity $v = c_0 = 1/\sqrt{\varepsilon_0 \mu_0}$, the above equations transform into Eq. (24.17) with the substitutions $\chi = \sqrt{\varepsilon_0/\mu_0}\, E_x$ and $\xi = H_y$. The *free-space impedance*, $\sqrt{\mu_0/\varepsilon_0} = 377\,\Omega$ in MKS units, equals the ratio between the magnitudes of the electric and magnetic fields in vacuum.

Since $\partial^2/\partial x\,\partial t = \partial^2/\partial t\,\partial x$, the wave equation can be recast in the form

$$\left(\frac{\partial^2}{\partial t^2} - v^2\frac{\partial^2}{\partial x^2}\right)\chi = \left(\frac{\partial}{\partial t} - v\frac{\partial}{\partial x}\right)\left(\frac{\partial}{\partial t} + v\frac{\partial}{\partial x}\right)\chi = 0 \tag{24.19}$$

In terms of

$$x_+ = \frac{x - vt}{2}, \qquad x_- = \frac{x + vt}{2} \tag{24.20}$$

for which $x = x_+ + x_-$ and $t = (x_- - x_+)/v$,

$$\frac{\partial}{\partial x_-} = \frac{\partial t}{\partial x_-}\frac{\partial}{\partial t} + \frac{\partial x}{\partial x_-}\frac{\partial}{\partial x} = \frac{1}{v}\frac{\partial}{\partial t} + \frac{\partial}{\partial x}$$

$$\frac{\partial}{\partial x_+} = \frac{\partial t}{\partial x_+}\frac{\partial}{\partial t} + \frac{\partial x}{\partial x_+}\frac{\partial}{\partial x} = -\frac{1}{v}\frac{\partial}{\partial t} + \frac{\partial}{\partial x} \tag{24.21}$$

Eq. (24.19) becomes

$$\frac{\partial^2}{\partial x_+\,\partial x_-}\chi(x_+, x_-) = 0 \tag{24.22}$$

The forward-traveling solution accordingly solves the *convection equation*

$$\frac{\partial\chi}{\partial x_-} = \frac{1}{v}\left(\frac{\partial}{\partial t} + v\frac{\partial}{\partial x}\right)\chi = 0 \tag{24.23}$$

yielding $\chi = \chi_F(x_+) = \chi_F(x - vt)$, while the general wave-equation solution is represented by $\chi = \chi_F(x - vt) + \chi_B(x + vt)$.

The operator $\partial/\partial t + v\,\partial/\partial x$, which corresponds to the change in a quantity measured in a reference system that moves at a velocity v, appears in numerous physical contexts. For example, in fluid mechanics, the mass of a fluid in a small volume of extent $\Delta V = \Delta x\,\Delta y\,\Delta z$ must be preserved over a small time interval Δt. Since the x-coordinate of a point in the fluid at x evolves over this time to the point $x + v_x(x)\Delta t$ while that of a point at $x + \Delta x$ evolves to $x + \Delta x + v_x(x + \Delta x)\Delta t \approx x + \Delta x + [(v_x(x) + \partial v_x/\partial x)\Delta x]\Delta t$, if $\rho(x, y, z)$ denotes the fluid density,

$$\rho\left(x + v_x\,\Delta t, y + v_y\,\Delta t, z + v_z\,\Delta t, t + \Delta t\right)$$

$$\times\left(\Delta x + \left(\frac{\Delta v_x}{\Delta x}\Delta x\right)\Delta t\right)\ldots\left(\Delta z + \left(\frac{\Delta v_z}{\Delta z}\Delta z\right)\Delta t\right)$$

$$= \rho(x, y, z, t)\Delta x\,\Delta y\,\Delta z \tag{24.24}$$

which yields the *equation of continuity* after expanding ρ on the left-hand side into a multidimensional Taylor series about (x, y, z, t) and retaining only first-order quantities,

$$\left\{\frac{\partial\rho}{\partial t} + \vec{v}\cdot\vec{\nabla}\rho + \rho\,\vec{\nabla}\cdot\vec{v}\right\}\Delta V\,\Delta t = 0 \tag{24.25}$$

In terms of the *convective* (*hydrodynamic or substantive*) derivative D/Dt,

$$\frac{D\rho}{Dt} = -\rho\,\vec{\nabla}\cdot\vec{v} \tag{24.26}$$

Hence, for example, if a liquid flows at 2.0 m/s in the z-direction and the fluid density is described by $\rho(z) = \rho_0 + c_z z + c_t t$, then in a frame moving with the fluid the change of the density with respect to time is $(\partial/\partial t + 2\,\partial/\partial z)(\rho_0 + c_z z + c_t t) = 2c_z + c_t$.

Similarly, the change in the velocity of a cubic fluid volume is determined by the imbalance of forces acting on the sides of the volume. Since the pressure, P, on the volume is defined to be positive acting inward, tracking a fluid element in the same manner as e.g. a thrown ball by moving along its path of evolution with its velocity $\vec{v}$, the velocity change of the element with time in the presence of external forces such as the gravitational field is given by (noting that the initial mass $\rho \Delta V$ of the volume is constant)

$$
\frac{\vec{v}\left(x + v_x\,\Delta t, y + v_y\,\Delta t, z + v_z\,\Delta t, t + \Delta t\right) - \vec{v}(x, y, z, t + \Delta t)}{\Delta t}
$$

$$
= \frac{\vec{F}}{\rho\,\Delta V}
$$

$$
= \frac{\rho\,\Delta V \vec{a}_{\text{external}}}{\rho\,\Delta V} + \frac{1}{\rho\,\Delta V}\left(\Delta x\,\Delta y\,\hat{e}_z\left[P_z\left(x + \frac{\Delta x}{2}, y + \frac{\Delta y}{2}, z\right)\right.\right.
$$

$$
\left.\left. - P_z\left(x + \frac{\Delta x}{2}, y + \frac{\Delta y}{2}, z + \Delta z\right)\right] + \cdots\right) \tag{24.27}
$$

Expanding in first-order quantities leads to *Euler's equation*

$$
\frac{D\vec{v}}{Dt} = \vec{a}_{\text{external}} - \frac{\vec{\nabla} P}{\rho} \tag{24.28}
$$

Hence for steady-state fluid flow ($\partial \vec{v}/\partial t = 0$) in a constant gravitational field, $\vec{a}_{\text{external}} = -g\hat{e}_z$, the velocity of an element of fluid falling in the $-\hat{e}_z$ direction with $\vec{v}(t = 0) = 0$ is given by $(\vec{v} \cdot \vec{\nabla})\vec{v} = (1/2)(\partial v^2/\partial z)\hat{e}_z = -g\hat{e}_z$, yielding $|v| = \sqrt{2g(-z)}$ as expected.

Discretizing the one-dimensional convection equation yields the *forward time centered space (FTCS)* procedure,

$$
\chi(x_i, t_{j+1}) = \chi(x_i, t_j) + \frac{v\,\Delta t}{2\,\Delta x}(\chi(x_{i+1}, t_j) - \chi(x_{i-1}, t_j))
$$

$$
\equiv \chi(x_i, t_j) + \frac{b}{2}(\chi(x_{i+1}, t_j) - \chi(x_{i-1}, t_j)) \tag{24.29}
$$

The quantity $b = v\Delta t/\Delta x$ is often termed the *Courant number*, since its value is related to the stability of a numerical procedure. However, the FTCS method is unstable for any Δt and Δx, as is apparent from the manner in which the initial field $\tilde{\chi}(x_i, t_0) = (0, 1, 0, -1, 0, 1\ldots)$ evolves over one propagation step. From Eq. (24.29), χ remains invariant for even-numbered points, whereas for odd-numbered points χ alternately increases and decreases by b, in both cases contributing a term proportional to b^2 to the norm of the field. A general mathematical treatment, the *Lax method*, evaluates the amplification of a specific term in the

Fourier-series expansion of the field by inserting

$$\chi(x_n, t_m) = e^{in\kappa \Delta x}\beta^m \tag{24.30}$$

into Eq. (24.29), resulting in

$$\beta = 1 + ib\sin(\kappa\,\Delta x) \tag{24.31}$$

and a corresponding amplification factor

$$|\beta| = \sqrt{1 + b^2\sin^2(\kappa\,\Delta x)} \geq 1 \tag{24.32}$$

This numerical instability is removed in the *Lax method* for small b by substituting the average $\big(\chi(x_{i-1}, t_j) + \chi(x_{i+1}, t_j)\big)/2$ for $\chi(x_i, t_j)$ on the right-hand side of Eq. (24.29):

$$\chi(x_i, t_{j+1}) = \frac{\chi(x_{i+1}, t_j) + \chi(x_{i-1}, t_j)}{2} + \frac{b}{2}\big(\chi(x_{i+1}, t_j) - \chi(x_{i-1}, t_j)\big) \tag{24.33}$$

This introduces a fictitious diffusion that couples into the field at each grid point for every time step a fraction of the field at the two neighboring grid points. Since the solutions of the diffusion equation decay with time, as observed above, the numerical diffusion counteracts the instability of the unmodified procedure. A Lax analysis yields in place of Eq. (24.32)

$$|\beta| = \sqrt{\cos^2(\kappa\,\Delta x) + b^2\sin^2(\kappa\,\Delta x)} \tag{24.34}$$

which is less than unity for $b < 1$, i.e. $\Delta t < \Delta x/v$. When $b > 1$, so that the field travels a distance greater than the distance between adjacent grid points in a single time step, the magnitude of the $\partial/\partial x$ convection term exceeds that of the diffusion term. Consequently the overall field behavior is dominated by convection, leading to amplification.

The *Lax–Wendroff* method introduces a diffusion term that is tailored to the specific analytic properties of the field. In particular, inserting the convection equation into the Taylor-series expansion

$$\begin{aligned}
\chi(x, t + \Delta t) &= \chi(x, t) + \Delta t\frac{\partial\chi}{\partial t} + \frac{(\Delta t)^2}{2}\frac{\partial^2\chi}{\partial t^2} + \cdots\\
&= \chi(x, t) - v\,\Delta t\frac{\partial\chi}{\partial x} + \frac{v^2(\Delta t)^2}{2}\frac{\partial^2\chi}{\partial x^2} + \cdots
\end{aligned} \tag{24.35}$$

yields the numerical procedure

$$\begin{aligned}
\chi(x_i, t_{j+1}) &= \chi(x_i, t_j) + \frac{b}{2}\big(\chi(x_{i+1}, t_j) - \chi(x_{i-1}, t_j)\big)\\
&\quad + \frac{b^2}{2}\big(\chi(x_{i+1}, t_j) - 2\chi(x_i, t_j) + \chi(x_{i-1}, t_j)\big)
\end{aligned} \tag{24.36}$$

In contrast to the Lax method, the diffusive coupling varies as $(\Delta t)^2$, while the numerical accuracy is enhanced.

The *hopscotch method*, known as the *Yee method* in electric-field simulation, solves Eq. (24.17) by placing the grid points for the field ξ equidistant from the points at which χ is evaluated while similarly employing a centered difference approximation to the time derivative. This results in

$$\xi(x_{i+1/2}, t_{j+1/2}) = \xi(x_{i+1/2}, t_{j-1/2}) + \frac{v\Delta t}{\Delta x} \left(\chi(x_{i+1}, t_j) - \chi(x_i, t_j) \right)$$

$$\chi(x_i, t_{j+1/2}) = \chi\left(x_i, t_{j-1/2}\right) + \frac{v\Delta t}{\Delta x} \left(\xi\left(x_{i+1/2}, t_j\right) - \xi\left(x_{i-1/2}, t_j\right) \right)$$

(24.37)

Applying the same procedure to the source-free curl equation in electromagnetics, $\partial \vec{B}/\partial t = -\vec{\nabla} \times \vec{E}$, yields for the time derivative of H_z the expression $\mu \partial H_z/\partial t = -(\partial E_x/\partial y - \partial E_y/\partial x)$, with a similar equation for the E field from $\partial \vec{D}/\partial t = \vec{\nabla} \times \vec{H}$. The z-component of the H-field equation can then be discretized as (for position-independent μ)

$$H_z\left(x_{i+\frac{1}{2}}, y_{j+\frac{1}{2}}, z_k, t_{m+\frac{1}{2}}\right) = H_z\left(x_{i+\frac{1}{2}}, y_{j+\frac{1}{2}}, z_k, t_{m-\frac{1}{2}}\right)$$

$$- \frac{\Delta t}{\mu} \left(\frac{E_y\left(x_{i+1}, y_{j+\frac{1}{2}}, z_k, t_m\right) - E_y\left(x_i, y_{j+\frac{1}{2}}, z_k, t_m\right)}{\Delta x} \right.$$

$$\left. - \frac{E_x\left(x_{i+\frac{1}{2}}, y_{j+1}, z_k, t_m\right) - E_x\left(x_{i+\frac{1}{2}}, y_j, z_k, t_m\right)}{\Delta y} \right)$$

(24.38)

The following Java program applies implements a one-dimensional hopscotch technique to a rightward-propagating exponential pulse:

```java
import de.dislin.*;
public class Hopscotch {
    static public void main( String[ ] args ) {
        int numberOfPoints = 200;
        int numberOfSteps = 500;
        float deltaX = 1.0F;
        float velocity = 1.0F;
        float deltaT = 1.0F;
        float field1[ ] = new float[numberOfPoints];
        float field2[ ] = new float[numberOfPoints];
        float xValues[ ] = new float[numberOfPoints];
        for ( int loop = 0; loop < numberOfPoints; loop++ ) {
            xValues[loop] = ( (float) loop ) * deltaX;
        }
        for ( int outerLoop = 0; outerLoop < numberOfSteps - 1;
            outerLoop++ ) {
        % Update the H field from the adjacent E field values
            for ( int loop = 0; loop < numberOfPoints - 1; loop++ )
                field2[loop] += velocity * deltaT *
                        ( field1[loop + 1] - field1[loop] ) / deltaX;
        % Update the E field from the adjacent H field values
            for ( int loop = 1; loop < numberOfPoints; loop++ )
                field1[loop] += velocity * deltaT *
                        ( field2[loop] - field2[loop - 1] ) / deltaX;
        % Specify the incoming electric field at the left boundary.
```

```
        field1[0] = (float) Math.exp( - ( outerLoop - 50 ) *
            ( outerLoop - 50 ) / 80. );
        if( outerLoop % 10 == 0 ) Dislin.qplot( xValues,
            field1, numberOfPoints );
    }
  }
}
```

Here fractional indices are replaced by integers. The excitation (the electric field in electromagnetic contexts), **field1**, enters the computational window through the first point of the grid. At the rightmost boundary, **field2** (the magnetic field), is set to zero and therefore reverses direction while the direction of **field1**, $\vec{E}$, remains fixed. Consequently $(\vec{E} \times \vec{B})$ reverses sign, changing the direction of propagation. When this reflected field reaches the left boundary, **field1** of the incoming excitation is negligible. The resulting zero boundary condition inverts the reflected electric field so that the left- and right-propagating electric fields cancel out at the boundary. Since the Courant number is unity in the above code, the field is displaced by one grid-point spacing for each time step.

Stability analyses can be applied to numerical procedures written in matrix form

$$\vec{\chi}(t_{j+1}) = \mathbf{D}\vec{\chi}(t_j) \tag{24.39}$$

in which the matrix $\mathbf{D}$ is obtained by replacing operators with matrices, as in

$$v\frac{\partial \chi}{\partial x} \rightarrow v \begin{pmatrix} 0 & 1 & & & & \\ -1 & 0 & 1 & & & \\ & -1 & 0 & 1 & & \\ & & & \ddots & & \\ & & & -1 & 0 & 1 \\ & & & & -1 & 0 \end{pmatrix} \begin{pmatrix} \chi(x_0, t_j) \\ \chi(x_1, t_j) \\ \chi(x_2, t_j) \\ \vdots \\ \chi(x_{N-2}, t_j) \\ \chi(x_{N-1}, t_j) \end{pmatrix} \tag{24.40}$$

Denoting the eigenvalues of $\mathbf{D}$ by λ_k, Eq. (24.39) implies that, for coefficients $a_k^{(j)}$, where the $a_k^{(0)}$ are determined by the initial conditions,

$$\vec{\chi}(t_{j+1}) = \sum_{k=1}^{N} a_k^{(j+1)}\varphi_k = \mathbf{D}\left(\sum_{k=1}^{N} a_k^{(j)}\varphi_k\right) = \sum_{k=1}^{N} a_k^{(j)}\lambda_k\varphi_k \tag{24.41}$$

After many time steps, the term or terms with the largest *spectral radius* $|\lambda_k|$ dominate and $\lambda_{\text{maximum}} = \lim_{j\to\infty} \mathbf{D}\vec{\chi}(t_j)/\vec{\chi}(t_j)$. This ratio can be evaluated at any spatial point x_i, or can be appropriately averaged over all points. If the largest eigenvalue is unique, the corresponding normalized eigenvector is given by $\lim_{j\to\infty} \vec{\chi}(t_j)/|\vec{\chi}(t_j)|$.

24.4 Elliptic equations

Elliptic equations are often solved by considering the associated diffusion problem in one additional dimension. For example, augmenting the two-dimensional

Poisson equation through the inclusion of an additional time dimension yields

$$\frac{\partial V(x,y,t)}{\partial t} = \frac{\partial^2 V(x,y,t)}{\partial x^2} + \frac{\partial^2 V(x,y,t)}{\partial y^2} + \frac{1}{\varepsilon_0}\rho(x,y) \qquad (24.42)$$

For time-independent boundary conditions, the solution to this problem converges to the steady-state solution for which $\partial V/\partial t$, and hence the right-hand side of the equation, equals zero given any initial conditions. Numerically, with $\Delta x = \Delta y$, this corresponds to solving repeatedly

$$V_{i,j}^{(n+1)} = \left(1 - \frac{4\,\Delta t}{(\Delta x)^2}\right) V_{i,j}^{(n)} + \frac{\Delta t}{(\Delta x)^2}\left(V_{i,j+1}^{(n)} + V_{i,j-1}^{(n)} + V_{i+1,j}^{(n)} + V_{i-1,j}^{(n)} + \frac{\rho(x,y)}{\varepsilon_0}\right) \qquad (24.43)$$

While V_{ij} here must be appropriately saved to avoid being overwritten before $V_{i,j-1}$ and $V_{i-1,j}$ are evaluated, a modified version of this procedure,

$$V_{i,j}^{(n+1)} = \left(1 - \frac{4\,\Delta t}{(\Delta x)^2}\right) V_{i,j}^{(n)} + \frac{\Delta t}{(\Delta x)^2}\left(V_{i,j+1}^{(n+1)} + V_{i,j-1}^{(n+1)} + V_{i+1,j}^{(n)} + V_{i-1,j}^{(n)} + \frac{\rho(x,y)}{\varepsilon_0}\right) \qquad (24.44)$$

eliminates this step. This procedure is termed the *Gauss–Seidel* method if $\Delta t/(\Delta x)^2 = 1/4$ and *underrelaxation* for $\Delta t/(\Delta x)^2 < 1/4$, and is unstable for $\Delta t/(\Delta x)^2 > 1/2$. For a rectangular grid and $\rho = 0$, the optimal value of $\Delta t/(\Delta x)^2 = 1/2(1 + \sqrt{1 - \gamma^2})$, with $2\gamma = \cos(\pi/N_x) + \cos(\pi/N_y)$.

For rectangular computational window boundaries, Poisson's equation can be solved directly by Fourier transforming. Since the FFT periodically extends its argument so that $V_0 = V_N$, if we denote by A_{l+1} the vector of A values shifted periodically by one grid position, then

$$[FFT(A_{l+1})]_m = \sum_{l=0}^{N-1} A_{l+1} e^{-i\frac{2\pi l m}{N}} = e^{i\frac{2\pi m}{N}} \sum_{l=0}^{N-1} A_{l+1} e^{-i\frac{2\pi m(l+1)}{N}}$$

$$= e^{i\frac{2\pi m}{N}} \sum_{l=1}^{N} A_l e^{-i\frac{2\pi l m}{N}} = e^{i\frac{2\pi m}{N}} \sum_{l=0}^{N-1} A_l e^{-i\frac{2\pi l m}{N}}$$

$$= e^{i\frac{2\pi m}{N}} [FFT(A_l)]_m \qquad (24.45)$$

Accordingly, if v_{mn} and r_{mn} are the Fourier components of V and ρ, respectively, fast Fourier transforming in two dimensions the discrete version of Poisson's equation (again for $\Delta x = \Delta y \equiv \Delta \alpha$)

$$V_{i,j+1}^{(n)} + V_{i,j-1}^{(n)} + V_{i+1,j}^{(n)} + V_{i-1,j}^{(n)} - 4V_{i,j}^{(n)} = -(\Delta \alpha)^2 \frac{\rho(x,y)}{\varepsilon_0} \qquad (24.46)$$

yields for periodic or, equivalently, zero boundary conditions

$$\left(e^{i\frac{2\pi m}{N}} + e^{-i\frac{2\pi m}{N}} + e^{i\frac{2\pi n}{N}} + e^{i\frac{2\pi n}{N}} - 4\right) v_{mn} = -(\Delta \alpha)^2 \frac{r_{mn}}{\varepsilon_0} \qquad (24.47)$$

or, equivalently,

$$v_{mn} = -\frac{(\Delta\alpha)^2 r_{mn}}{2\left(\cos\left(2\pi m/N\right) + \cos\left(2\pi n/N\right) - 2\right)} \tag{24.48}$$

after which $V = \text{IFFT}(v)$ again in two dimensions.

24.5 Split-operator methods for parabolic differential equations

To evolve a complex quantum-mechanical wavefunction, $\psi(x, t)$, in time numerically according to the one-dimensional Schrödinger equation,

$$\frac{d\psi}{dt} = -\frac{i}{\hbar}\left(-\frac{\hbar^2}{2m}\frac{\partial^2}{\partial x^2} + V(x)\right)\psi \equiv -\frac{i}{\hbar}H\psi = -\frac{i}{\hbar}(T+V)\psi \tag{24.49}$$

note that the field evolution for a sufficiently small time step is described by

$$\psi(t + \Delta t) \approx \left(1 - \frac{i}{\hbar}H\,\Delta t\right)\psi(t) \tag{24.50}$$

In a finite-difference discretization, Eq. (24.50) adopts the form

$$|\psi(t + \Delta t)\rangle = \left(\mathbf{I} - \frac{i\,\Delta t}{\hbar}\mathbf{H}\right)|\psi(t)\rangle \equiv \mathbf{U}(\Delta t)|\psi(t)\rangle$$

which represents

$$\begin{pmatrix} \psi(x_1, t+\Delta t) \\ \psi(x_2, t+\Delta t) \\ \psi(x_3, t+\Delta t) \\ \vdots \\ \psi(x_N, t+\Delta t) \end{pmatrix} = \begin{pmatrix} \psi(x_1, t) \\ \psi(x_2, t) \\ \psi(x_3, t) \\ \vdots \\ \psi(x_N, t) \end{pmatrix} - \frac{i\,\Delta t}{\hbar}\mathbf{H} \begin{pmatrix} \psi(x_1, t) \\ \psi(x_2, t) \\ \psi(x_3, t) \\ \vdots \\ \psi(x_N, t) \end{pmatrix} \tag{24.51}$$

In one dimension with zero boundary conditions,

$$\mathbf{H} = \frac{-\hbar^2}{2m(\Delta x)^2}\begin{pmatrix} -2 & 1 & & \\ 1 & -2 & \ddots & \\ & \ddots & \ddots & 1 \\ & & 1 & -2 \end{pmatrix} + \begin{pmatrix} V(x_1) & & & \\ & V(x_2) & & \\ & & \ddots & \\ & & & V(x_N) \end{pmatrix} \tag{24.52}$$

For a confined, non-decaying particle, the time evolution operator $\mathbf{U}(\Delta t)$ must conserve the total probability of the particle within its containing volume, $[x_1, x_n]$. This probability equals the sum of the probabilities (squared wavefunctions) of finding the particle at each point within the volume, so that

$$\langle\psi|\psi\rangle \equiv \left(\psi^*(x_1, t), \psi^*(x_2, t), \psi^*(x_3, t), \ldots, \psi^*(x_N, t)\right)\begin{pmatrix} \psi(x_1, t) \\ \psi(x_2, t) \\ \psi(x_3, t) \\ \vdots \\ \psi(x_N, t) \end{pmatrix}$$

$$= |\psi(x_1, t)|^2 + |\psi(x_2, t)|^2 + \cdots + |\psi(x_N, t)|^2 \tag{24.53}$$

must be preserved under the transformation $\psi(t + \Delta t) = U\varphi(t)$. That is, noting that $(\mathbf{AB})^\dagger = \mathbf{B}^\dagger\mathbf{A}^\dagger$, this requires

$$\langle\psi(t+\Delta t)|\psi(t+\Delta t)\rangle = \langle\mathbf{U}\psi(t)|\mathbf{U}\psi(t)\rangle = \langle\psi(t)|\mathbf{U}^\dagger\mathbf{U}|\psi(t)\rangle = \langle\psi(t)|\psi(t)\rangle \quad (24.54)$$

Hence $\mathbf{U}^\dagger\mathbf{U} = \mathbf{I}$, or

$$\mathbf{U}^\dagger = \mathbf{U}^{-1} \quad (24.55)$$

i.e. $\mathbf{U}$ is *unitary*. Since $\mathbf{U}(\Delta t) = \mathbf{I} - i\,\Delta t\mathbf{H}/\hbar$, this results in the condition that

$$\mathbf{U}^\dagger(\Delta t)\mathbf{U}(\Delta t) = \left(\mathbf{I} + \frac{i\,\Delta t}{\hbar}\mathbf{H}^\dagger\right)\left(\mathbf{I} - \frac{i\,\Delta t}{\hbar}\mathbf{H}\right)$$

$$= \mathbf{I} - \frac{i\,\Delta t}{\hbar}\left(\mathbf{H} - \mathbf{H}^\dagger\right) - \left(\frac{\Delta t}{\hbar}\right)^2\mathbf{H}^\dagger\mathbf{H} \quad (24.56)$$

approaches $\mathbf{I}$ to second order as $\Delta t \to 0$. Accordingly, $\mathbf{H} = \mathbf{H}^\dagger$, so that $\mathbf{H}$ must be Hermitian or, equivalently, $\mathbf{S} = i\mathbf{H}$ is *skew-Hermitian* with the property $\mathbf{S}^\dagger = -\mathbf{S}$.

The evolution operator over a *non-infinitesimal* time interval is then, for a time-independent potential function,

$$\mathbf{U}(t) = \lim_{N\to\infty}\left(1 - \frac{it}{\hbar N}\mathbf{H}\right)^N$$

$$= \lim_{N\to\infty}\left(1 - \frac{N}{1!}\frac{it}{\hbar N}\mathbf{H} + \frac{N(N-1)}{2!}\left(\frac{it}{\hbar N}\right)^2\mathbf{H}^2 \mid \cdots\right) - e^{-i\frac{t}{\hbar}\mathbf{H}} \quad (24.57)$$

which is unitary with inverse operator

$$\mathbf{U}^{-1}(t) = \mathbf{U}(-t) = e^{i\frac{t}{\hbar}\mathbf{H}} = \mathbf{U}^\dagger(t)$$

as can be explicitly verified by taking the Hermitian conjugate of the individual terms in the Taylor-series expansion of Eq. (24.57) and employing the Hermiticity of $\mathbf{H}$.

That $\mathbf{U}(t)$ conserves the square of the field can also be verified by first recalling that all eigenvalues of a Hermitian matrix are real since, if φ_i denotes an eigenfunction of $\mathbf{H}$ such that $\mathbf{H}\varphi_i = E_i\varphi_i$,

$$E = \langle\varphi_i|\mathbf{H}|\varphi_i\rangle = \langle\varphi_i|\mathbf{H}^\dagger|\varphi_i\rangle = (\mathbf{H}|\varphi_i\rangle)^\dagger|\varphi_i\rangle = (E|\varphi_i\rangle)^\dagger|\varphi_i\rangle = E^*\langle\varphi_i|\varphi_i\rangle = E^* \quad (24.58)$$

Accordingly, since any wavefunction, ψ, can be expressed as a superposition, $|\psi\rangle = \sum_{m=1}^N c_m|\varphi_m\rangle$, of the orthonomalized eigenfunctions of $\mathbf{H}$,

$$e^{-i\frac{t}{\hbar}\mathbf{H}}|\varphi\rangle = e^{-i\frac{t}{\hbar}\mathbf{H}}\left(\sum_{m=1}^N c_m|\varphi_m\rangle\right) = \sum_{m=1}^N c_m e^{-i\frac{t}{\hbar}E_m}|\varphi_m\rangle \quad (24.59)$$

However,

$$\left(e^{-i\frac{t}{\hbar}\mathbf{H}}|\varphi\rangle \right)^{\dagger} e^{-i\frac{t}{\hbar}\mathbf{H}}|\varphi\rangle = \left(\sum_{n=1}^{N} \langle \varphi_n | c_n^* e^{i\frac{t}{\hbar}E_n} \right) \left(\sum_{m=1}^{N} c_m e^{-i\frac{t}{\hbar}E_m} |\varphi_m\rangle \right)$$

$$= \sum_{n,m=1}^{N} c_n^* c_m e^{-i\frac{t}{\hbar}(E_m - E_n)} \underbrace{\langle \varphi_n \mid \varphi_m \rangle}_{\delta_{nm}}$$

$$= \sum_{m=1}^{N} c_m^* c_m = 1 \tag{24.60}$$

Hence, $e^{-i\frac{t}{\hbar}\mathbf{H}}$ preserves the norm of a propagating field and is therefore unitary. Numerical procedures, which are typically derived from approximations to

$$\varphi(t + \Delta t) = e^{-\frac{i\Delta t}{\hbar}(T+V)}\varphi(t) \tag{24.61}$$

should consequently preserve this property.

The difficulty encountered in evaluating Eq. (24.61) arises from the non-commutativity of $T = T(p) = T(-i\hbar\,\vec{\nabla})$ and $V = V(\vec{r})$. Generally, however, if A and B do not commute, while α is a small quantity,

$$e^{\alpha(A+B)} \approx 1 + \alpha(A + B) + \frac{\alpha^2}{2!}\left(A^2 + AB + BA + B^2\right) + O(\alpha^3)$$

$$\approx \left(1 + \frac{\alpha}{2}A + \frac{\alpha^2 A^2}{4}\right)\left(1 + \alpha B + \frac{\alpha^2 B^2}{2}\right)\left(1 + \frac{\alpha}{2}A + \frac{\alpha^2 A^2}{4}\right) + O(\alpha^3)$$

$$\approx e^{\alpha\frac{A}{2}} e^{\alpha B} e^{\alpha\frac{A}{2}} + O(\alpha^3) \equiv \Lambda_3(\alpha) + O(\alpha^3) \tag{24.62}$$

while higher-order accurate product expansions can be obtained by noting e.g. that

$$e^{\alpha(A+B)} = \Lambda_3(x\alpha)\Lambda_3\left((1-2x)\alpha\right)\Lambda_3(x\alpha)$$

$$+ \left(2x^3 + (1-2x)^3\right)O(\alpha^3) + \underbrace{O(\alpha^4)}_{O(\alpha)^5 \text{ by symmetry}} \tag{24.63}$$

Solving $2x^3 + (1 - 2x)^3 = 0$ thus yields a fifth-order-accurate expression. Such *operator-splitting methods* form the basis of numerous solution techniques, since the exponentials of the individual operators T and V are much more simply evaluated than is the exponential of their sum, Eq. (24.61). For example, one procedure, the *split-step fast Fourier transform method* evaluates the operator $\exp(i\hbar\,\Delta t(\partial^2/\partial x^2)/(2m))$ after fast Fourier transforming according to

$$e^{\frac{i\hbar\Delta t}{2m}\frac{\partial^2}{\partial x^2}}\psi = \text{IFT}\left(e^{-\frac{i\hbar\Delta t}{2m\hbar}k_m^2}[\text{FT}(\psi)]_m\right) \tag{24.64}$$

as discussed in detail in Section 24.7.

24.6 Symplectic evolution operators in classical mechanics

Analogous split-operator procedures in mechanics can be obtained from Hamilton's equations

$$\frac{dx_m}{dt} = \frac{\partial H}{\partial p_m}$$
$$\frac{dp_m}{dt} = -\frac{\partial H}{\partial x_m}$$

24.65)

which imply that, if the Hamiltonian H is a function of N "generalized" position and momentum variables, then, for any function $f(x_m, p_m, t)$ of these variables,

$$\frac{df}{dt} = \frac{\partial f}{\partial t} + \sum_{m=1}^{N} \left(\frac{\partial f}{\partial x_m} \frac{dx_m}{dt} + \frac{\partial f}{\partial p_m} \frac{dp_m}{dt} \right)$$

$$= \frac{\partial f}{\partial t} + \sum_{m=1}^{N} \left(\frac{\partial f}{\partial x_m} \frac{\partial H}{\partial p_m} - \frac{\partial f}{\partial p_m} \frac{\partial H}{\partial x_m} \right)$$

$$\equiv \frac{\partial f}{\partial t} + \{f, H\} \equiv \frac{\partial f}{\partial t} + D_H f \qquad (24.66)$$

where $\{\ \}$ is termed the *Poisson bracket*. If f depends only on position and momentum, then $\partial f / \partial t = 0$ and

$$f(t + \Delta t) = e^{\Delta t D_H} f(t) \qquad (24.67)$$

which is termed a *symplectic* propagation method.

On specializing again to one dimension, assuming that $H = T(p) + V(x)$ and, finally, combining the position and momentum variables into a vector, the above equation can be approximated as

$$\begin{pmatrix} x(t + \Delta t) \\ p(t + \Delta t) \end{pmatrix} = e^{\Delta t D_{T+V}} \begin{pmatrix} x(t) \\ p(t) \end{pmatrix} \approx e^{\frac{\Delta t}{2} D_T} e^{\Delta t D_V} e^{\frac{\Delta t}{2} D_T} \begin{pmatrix} x(t) \\ p(t) \end{pmatrix} + O(\Delta t)^3 \quad (24.68)$$

in which e.g.

$$e^{\Delta t D_T} \begin{pmatrix} x(t) \\ p(t) \end{pmatrix} = \left(\sum_{m=0}^{\infty} (\Delta t \, D_T)^m \right) \begin{pmatrix} x(t) \\ p(t) \end{pmatrix} \qquad (24.69)$$

However,

$$D_T \begin{pmatrix} x \\ p \end{pmatrix} = \left\{ \begin{pmatrix} x \\ p \end{pmatrix}, T(p) \right\} = \begin{pmatrix} \dfrac{\partial x}{\partial x} \dfrac{\partial T(p)}{\partial p} - \dfrac{\partial x}{\partial p} \dfrac{\partial T(p)}{\partial x} \\ \dfrac{\partial p}{\partial x} \dfrac{\partial T(p)}{\partial p} - \dfrac{\partial p}{\partial p} \dfrac{\partial T(p)}{\partial x} \end{pmatrix} = \begin{pmatrix} \dfrac{\partial T(p)}{\partial p} \\ 0 \end{pmatrix} \quad (24.70)$$

while

$$D_T^2 \begin{pmatrix} x \\ p \end{pmatrix} = \left\{ \begin{pmatrix} \dfrac{\partial T(p)}{\partial p} \\ 0 \end{pmatrix}, T(p) \right\} = 0 \qquad (24.71)$$

Hence,

$$e^{\frac{\Delta t}{2} D_T} \begin{pmatrix} x(t) \\ p(t) \end{pmatrix} = \begin{pmatrix} x(t) + \dfrac{\Delta t}{2} \dfrac{\partial T(p)}{\partial p} \\ p(t) \end{pmatrix} \tag{24.72}$$

Similarly,

$$e^{\Delta t D_V} \begin{pmatrix} x(t) \\ p(t) \end{pmatrix} = \begin{pmatrix} x(t) \\ p - \Delta t \dfrac{\partial V(x)}{\partial x} \end{pmatrix} \tag{24.73}$$

If $T(p) = p^2/(2m)$, this corresponds to first updating $x(t + \Delta t/2) = x(t) + \Delta t v(t)/2$ and then applying $p(t + \Delta t) = p(t) + \Delta t\, F(x(t + \Delta t/2))$ followed by a subsequent displacement step. Higher-order operator product expansions evaluate the quantities in Eqs. (24.72) and (24.73) at non-standard distances and times.

24.7 Fast Fourier transform methods in optics

As noted above, split-operator methods can be unified with fast Fourier transform procedures that accurately represent derivative operators. However, the fast Fourier transform implicitly implements periodic boundary conditions. Radiation boundary conditions that remove the power incident on the computational window edges are therefore often simulated by introducing an artificial absorbing region into the potential function near the window boundaries. While this manifestly removes the unitarity of the procedure, since the power in the field monotonically decreases, numerical divergences are suppressed rather than amplified.

To illustrate the method, consider the hyperbolic scalar wave equation that describes the evolution of a monochromatic (single-frequency) electric field with time when polarization effects can be neglected:

$$\frac{\partial^2 E}{\partial x^2} + \frac{\partial^2 E}{\partial y^2} + \frac{\partial^2 E}{\partial z^2} + k_0^2 n^2(x, y, z)E = 0 \tag{24.74}$$

The wavenumber $k_0 = 2\pi/\lambda_0$, where λ_0 is the wavelength of the electric field in vacuum and n denotes the refractive index. If the dominant field components propagate close to an angle of θ_0 with respect to the z-axis in a region with average refractive index $\bar{n}$, a "reference refractive index" given by $n_0 = \bar{n} \cos\theta_0$ can be introduced. Then, with

$$X_0 = \frac{1}{n_0^2 k_0^2} \frac{\partial^2}{\partial x^2}, \qquad Y_0 = \frac{1}{n_0^2 k_0^2} \frac{\partial^2}{\partial y^2} \tag{24.75}$$

and

$$N = \frac{n^2}{n_0^2} - 1 \tag{24.76}$$

Eq. (24.74) adopts the form

$$\frac{\partial^2 E}{\partial z^2} + k_0^2 n_0^2 (1 + X_0 + Y_0 + N)E = 0 \tag{24.77}$$

This equation is of second order in the longitudinal variable z and therefore possesses both forward- and backward-traveling solutions. However, if N varies sufficiently slowly in the z-direction that reflection effects are negligible, the forward- and backward-traveling fields can again be isolated by formally factoring Eq. (24.77),

$$\left(\frac{\partial}{\partial z} - \delta\sqrt{1 + X_0 + Y_0 + N}\right)\left(\frac{\partial}{\partial z} + \delta\sqrt{1 + X_0 + Y_0 + N}\right) E = 0 \qquad (24.78)$$

with $\delta = -ik_0n_0$. Following the convention that the time dependence of the electric field is given by $\exp(i\omega t)$, the forward-propagating field is then identified with the solution of

$$\left(\frac{\partial}{\partial z} - \delta\sqrt{1 + X_0 + Y_0 + N}\right) E = 0 \qquad (24.79)$$

The *Fresnel equation* is obtained if we assume that $(X_0 + Y_0 + N)E \ll E$ and replace the square root by the first-order term in its Taylor-series expansion:

$$\frac{\partial E}{\partial z} - \delta\left[1 + \frac{1}{2}(X_0 + Y_0 + N)\right] E = 0 \qquad (24.80)$$

After the rapidly varying component of the electric field has been removed by introducing the modified field

$$E(x, y, z) = E(x, y, z)e^{\delta} = E(x, y, z)e^{-in_0k_0 z} \qquad (24.81)$$

the split-operator method yields

$$E(x, y, z + \Delta z) = e^{-\frac{i\Delta z}{4n_0k_0}\left(\frac{\partial^2}{\partial x^2} + \frac{\partial^2}{\partial y^2}\right)} e^{-\frac{ik_0\Delta z}{2n_0}\left(n^2 - n_0^2\right)} e^{-\frac{i\Delta z}{4n_0k_0}\left(\frac{\partial^2}{\partial x^2} + \frac{\partial^2}{\partial y^2}\right)} E(x, y, z) + O(\Delta z)^3 \qquad (24.82)$$

Note that, for $n = n_0 + \Delta n$, so that $n^2 - n_0^2 \approx 2\,\Delta n\, n_0$, the exponent of the second exponential operator is simply $-ik_0\,\Delta n\,\Delta z$, which is $-i$ times the approximate phase change ($\Delta\phi = k\,\Delta x$) of light resulting from the refractive-index inhomogeneity.

To apply Eq. (24.82), the field and refractive index are first discretized on an equidistant spatial grid

$$\left\{(x_i, y_j) : x_i = -L_x\left(1 - \frac{2(i-1)}{N_x}\right), y_j = -L_y\left(1 - \frac{2(j-1)}{N_y}\right),\right.$$

$$\left. i = 1, \ldots, N_x, \; j = 1, \ldots, N_y\right\} \qquad (24.83)$$

The length L_α with $\alpha = x, y$ coincides with the total distance covered by N intervals between adjacent points. The first and third operators in Eq. (24.82) are evaluated after application of the FFT to obtain the Fourier coefficients $E_{mn}(z)$. Subsequently, the inverse Fourier transform is employed according to

$$E(x_k, y_l, z) = \sum_{m=-\frac{N_x}{2}+1}^{\frac{N_x}{2}} \sum_{n=-\frac{N_y}{2}+1}^{\frac{N_y}{2}} E_{mn}(z)\, e^{\frac{i\Delta z}{4n_0 k_0}\left(\left(\frac{2\pi m}{L_x}\right)^2+\left(\frac{2\pi n}{L_y}\right)^2\right)}\, e^{2\pi i\left(\frac{mx_k}{L_x}+\frac{ny_l}{L_y}\right)} \quad (24.84)$$

In the normal representation of the FFT the negative m and n values in the above expression are expressed as positive quantities $m + N, n + N$.

The unitarity of the above propagation algorithm insures that radiated power remains within the computational window. Indeed, since the fast Fourier transform is an expansion in functions that are periodic with respect to the window length, the propagation method solves the problem for which the electric field and refractive index are periodically continued outside the window boundary. Thus, the field exiting the right side of the computational window simultaneously reappears at the left side of the window transported by its periodic extension. This effect can be suppressed only by introducing non-unitarity in the form of a complex (absorptive) component refractive index near the window boundaries, typically varying as

$$\frac{1}{2}\left[1 + \cos\left(\frac{\pi|x - x_\alpha|}{w}\right)\right] \quad |x - x_\alpha| < w$$
$$0 \qquad\qquad\qquad\qquad \text{elsewhere} \quad (24.85)$$

for each transverse direction x, where $x - x_\alpha$ is zero at the boundary position and the width of the absorbing region $w \ll L_\alpha$.

In the following program, a localized, Gaussian electric field distribution is first propagated 500 μm in vacuum with the split-operator fast Fourier transform method, then through a thin lens with a focal length of 250 μm and finally again a further 500 μm through vacuum to the corresponding image point. The thin lens is, for simplicity, here modeled by multiplication by a single operator that compensates for the unequal path lengths of rays propagating at different angles to its symmetry axis. In a more comprehensive analysis, a half propagation step would be employed at the beginning and end of the calculation as well as just before and just after the lens. Additionally, the exact spatially varying refractive index of the lens can be introduced and the field advanced through the lens in numerous small steps according to Eq. (24.82).

To simplify the lens operator, recall that for a thin lens $1/p + 1/q = 1/f$, where p and q correspond to the object and image lengths. For $p = q = 2f$, the phase change of a light beam that travels from the object point to a point in the lens a distance r from the symmetry axis and back to the image point is

$$2k_0 \sqrt{r^2 + 4f^2} + \Delta\phi(r) \approx 2k_0 \left(2f + r^2/(4f)\right) + \Delta\phi(r) \qquad (24.86)$$

where $\Delta\phi(r)$ is the phase change in the lens. For Eq. (24.86) to be independent of r requires

$$\Delta\phi(r) = c - \frac{k_0 r^2}{2f} \qquad (24.87)$$

Since a constant phase change does not affect optical field propagation, in the program below the central operator of Eq. (24.82) is replaced by $-i k_0 r^2/(2f)$:

```
vacuumWaveLength = 1.0;
gridLength = 200;
numberOfGridPoints = 1024;      % Must be even (preferably 2^n)
numberOfSpatialSteps = 20;      % Total number of propagation steps
propagationDistance = 1000;     % Total propagation distance
referenceIndex = 1.0;
propagationStepLength = propagationDistance / numberOfSpatialSteps;
pointLocationsPlus1R = linspace( -gridLength / 2, ...
      gridLength / 2, numberOfGridPoints + 1 );
pointLocationsR = pointLocationsPlus1R( 1 : numberOfGridPoints );
k0 = 2 * pi / vacuumWaveLength;      % Vacuum wavevector
% Gaussian initial field
fieldR = exp( -pointLocationsR.^2 / ( 2. * 5.0^2 ) );
% Second derivative operatior in Fourier transform representation
squaredFourierWavevectorComponentsR = ...
      -( 2 * pi / gridLength )^2 * [ 0 : numberOfGridPoints / 2, ...
      -numberOfGridPoints / 2 + 1 : -1 ].^2;
% Propagation operator for one propagation step
propagationOperatorR = exp( - i * propagationStepLength / ...
      ( 2 * k0 * referenceIndex ) * ...
      squaredFourierWavevectorComponentsR );
% Operator for the phase change in a thin lens
lensOperatorR = exp( i * k0 * pointLocationsR.^2 / 500 );
for loop = 1 : numberOfSpatialSteps
% A single propagation step is employed at the
% beginning instead of a half-step
      fieldR = ifft( propagationOperatorR .* fft( fieldR ) );
% The lens operator is applied near the
% middle of the propagation distance
      if loop == floor( numberOfSpatialSteps / 2 )
            fieldR = lensOperatorR .* fieldR;
      end
      plot( pointLocationsR, abs( fieldR ) )
      drawnow
end
```

The accuracy with which derivatives are evaluated with fast Fourier transform methods particularly favors the analysis of non-linear equations typified by the *non-linear Schrödinger equation*

$$i\frac{\partial E}{\partial t} + \frac{1}{2}\frac{\partial^2 E}{\partial x^2} + |E|^2 E = 0 \qquad (24.88)$$

which supports the stationary "soliton" solutions

$$E = ae^{-i\left(vx + (v - a^2)\frac{t}{2}\right)} \operatorname{sech}(a(x + vx)) \qquad (24.89)$$

that balance the dispersion (broadening) induced by the coupling of the field at adjacent spatial points through the second derivative term with the spatial focusing toward regions of large $|E|^2$ supplied by the non-linear term.

A program that propagates a soliton of unit amplitude is given below. The soliton exits the right-hand side of the computational window and reenters the left-hand side as a result of the periodic boundary conditions implicit in the FFT. This behavior is often exploited to examine multiple soliton collisions.

```
gridLength = 50;
numberOfGridPoints = 512;                % Must be even (preferably 2^n)
numberOfTimeSteps = 500;
propagationTime = 8;
propagationStepTime = propagationTime / numberOfTimeSteps;
pointLocationsPlus1R = linspace( -gridLength / 2, ...
    gridLength / 2, numberOfGridPoints + 1 );
pointLocationsR = pointLocationsPlus1R( 1 : numberOfGridPoints );
% Soliton initial field
fieldR = 1.0 ./ cosh( pointLocationsR - 8. ) .* ...
    exp( -i * 2 * pi * sin( 50 * pi / 180. ) * pointLocationsR );
% Free-space propagation operator for a half time step
propagationOperatorR = exp( -i * propagationStepTime * ...
    ( 2 * pi / gridLength )^2 / 4 * ...
    [ 0 : numberOfGridPoints / 2 , -numberOfGridPoints / 2 ...
    + 1 : -1 ] .^2 );
for loop = 1 : numberOfTimeSteps
    fieldR = ifft( propagationOperatorR .* ...
    fft( exp( i * propagationStepTime * conj( fieldR ) .* ...
        fieldR ) .* ifft( propagationOperatorR .* fft( fieldR ) ) ) );
    if rem( loop - 1, 50 ) == 0
        plot( pointLocationsR, abs( fieldR ) )
        drawnow
    end
end
```

Finally, the fast Fourier transform method is applied to the one-dimensional time-dependent Schrödinger equation for the propagation of a Gaussian wavepacket in a square-well potential of depth and width 5.0 a.u. (atomic units, for which $\hbar = m = e = 1$) in the presence of zero boundary conditions. The result is compared with the analytic expression for $V(x) = 0$. (The negative sign in the propagation operator results from the product of three – signs, one in the kinetic-energy term $T = -(\hbar^2/(2m))\partial/\partial x^2$, the second from multiplication of both sides of the Schrödinger equation by $-i$ to remove the $+i$ from the time derivative, and the third from $k_m^2 \to -(2\pi m/L)^2$ in the discrete Fourier representation.

All Schrödinger-equation propagation methods below require the following square-well potential function (normally **wellWidth** and **wellDepth** would be passed to the function from the calling program either through a **global** statement or as function parameters). A more accurate treatment insures that the neighboring grid points are spaced equidistantly from the edges of the square-well potential.

```
file potential.m
function y = potential( x )
wellWidth = 5;
wellDepth = 5;
y = zeros( length( x ), 1 );
for loop = 1 : length( x )
    if abs( x(loop) ) < wellWidth;
        y(loop) = -wellDepth;
    end
end
```

```
file schfft.m
hold on;
stepLength = 0.005;
numberOfTimeSteps = 500;
fieldWidth = 2;
computationalWindowWidth = 20;
numberOfPoints = 300;                 % Must be even (preferably 2^n)
gridPointsPlus1R = linspace( -computationalWindowWidth / 2, ...
    computationalWindowWidth / 2, numberOfPoints + 1 );
gridPointsC = gridPointsPlus1R( 1 : numberOfPoints ).';
wavefunctionC = exp( -gridPointsC.^2 ./ ( 2 * fieldWidth^2 ) );
% Kinetic-energy part of propagation operator
propagationOperatorC = exp( -i * stepLength / 2 * ...
    ( 2 * pi / computationalWindowWidth )^2 * ...
    ( [ 0 : numberOfPoints / 2 , ...
    -numberOfPoints / 2 + 1 : -1 ]').^2 );
% Potential-energy part of propagation operator
potentialOperatorC = exp( -i * stepLength * ...
     potential( gridPointsC ) );
for loop = 1 : numberOfTimeSteps
% Fourier transform method wavefunction propagation
    wavefunctionC = ifft( propagationOperatorC .* ...
        fft( potentialOperatorC .* wavefunctionC ) );
    if ( rem(loop, 50) == 0 )
        plot( gridPointsC, abs( wavefunctionC ), 'r' );
% Analytic expression for Gaussian wavepacket
% propagation in constant potential
        coefficient = ( 1 + ( stepLength * loop )^2 / ...
          fieldWidth^4 );
        plot( gridPointsC, coefficient^-0.25 * ...
```

```
            exp ( -gridPointsC.^2 ./ ...
            ( 2 * fieldWidth^2 * coefficient ) ), 'k' );
    end
    drawnow
end
```

24.8 The Crank–Nicholson method in quantum mechanics

A Padé approximant to a function generally is applicable in a wider domain than the Taylor-series expansions of equivalent order since it contains higher-order power-series terms that approximate those of the function. For example, comparing the Taylor-series expansion of the exponential function and its $(1,1)$ Padé-series approximation,

$$
e^x = 1 + x + \frac{x^2}{2} + \frac{x^3}{6} + \frac{x^4}{24} + \cdots
$$
$$
\frac{1 + x/2}{1 - x/2} = 1 + x + \frac{x^2}{2} + \frac{x^3}{4} + \frac{x^4}{8} + \cdots
$$

(24.90)

indicates that the Padé approximant approximates properties of the higher-order terms, in contrast to a two-term Taylor-series truncation of the power series. Applying this technique to the exponential propagation operator yields

$$
e^{-\frac{i \Delta t}{\hbar} \mathbf{H}} \approx \left(1 + \frac{i \, \Delta t}{2 \hbar} \mathbf{H} \right)^{-1} \left(1 - \frac{i \, \Delta t}{2 \hbar} \mathbf{H} \right)
$$

(24.91)

This operator is unitary since

$$
\left(1 + \frac{i \, \Delta t}{2 \hbar} \mathbf{H} \right)^{-1} \left(1 - \frac{i \, \Delta t}{2 \hbar} \mathbf{H} \right) |\psi\rangle = \sum_{m=1}^{N} c_m \frac{1 - \dfrac{i \, \Delta t}{2 \hbar} E_m}{1 + \dfrac{i \, \Delta t}{2 \hbar} E_m} |\varphi_m\rangle
$$

(24.92)

for which the coefficient of each eigenfunction has magnitude $|c_m|$ as in Eq. (24.60).

The *Crank–Nicholson* procedure is implemented as follows:

$$
\begin{aligned}
\psi(t + \Delta t) &= \left(1 + \frac{i \, \Delta t}{2 \hbar} \mathbf{H} \right)^{-1} \left(1 - \frac{i \, \Delta t}{2 \hbar} \mathbf{H} \right) \psi(t) \\
&= \left(1 + \frac{i \, \Delta t}{2 \hbar} \mathbf{H} \right)^{-1} \left(- \left(1 + \frac{i \, \Delta t}{2 \hbar} \mathbf{H} \right) + 2 \right) \psi(t) \\
&= \left(-1 + 2 \left(1 + \frac{i \, \Delta t}{2 \hbar} \mathbf{H} \right)^{-1} \right) \psi(t) \\
&= -\psi(t) + 2 \chi
\end{aligned}
$$

(24.93)

where

$$
\left(1 + \frac{i \, \Delta t}{2 \hbar} \mathbf{H} \right) \chi = \psi(t)
$$

(24.94)

Accordingly, each time step requires the solution of a tridiagonal equation system. In multiple dimensions the commutativity of e.g. $\partial^2/\partial x^2$ and $\partial^2/\partial y^2$ further enables the *alternating directional method* (ADI),

$$e^{-\frac{i\Delta t}{\hbar}T} = e^{\frac{i\hbar}{2m}\left(\frac{\partial^2}{\partial x^2} + \frac{\partial^2}{\partial y^2} + \frac{\partial^2}{\partial z^2}\right)} = e^{\frac{i\hbar}{2m}\frac{\partial^2}{\partial x^2}} e^{\frac{i\hbar}{2m}\frac{\partial^2}{\partial y^2}} e^{\frac{i\hbar}{2m}\frac{\partial^2}{\partial z^2}} \tag{24.95}$$

Each exponential operator can then be separately evaluated with the Crank–Nicholson procedure.

The following Octave program applies the Crank–Nicholson method to the one-dimensional time-dependent Schrödinger equation for the propagation of a Gaussian wavepacket in a square-well potential of depth and width 5.0 a.u. (atomic units, for which $\hbar = m = e = 1$) in the presence of zero boundary conditions. The result is then compared with the analytic expression for $V(x) = 0$:

```
hold on;
clear all;
stepLength = .005;
numberOfTimeSteps = 500;
fieldWidth = 2;
computationalWindowWidth = 20;
numberOfPoints = 300;
dx = computationalWindowWidth / ( numberOfPoints - 1 );
gridPointsC = linspace( -computationalWindowWidth / 2, ...
    computationalWindowWidth / 2, numberOfPoints ).';
wavefunctionC = exp( -gridPointsC.^2 ./ ( 2 * fieldWidth^2 ) );

% Crank-Nicolson solver based on user-supplied
% tridiagonal equation solution routine
enableMyTridiagonal = 1;
if enableMyTridiagonal
    aC = stepLength/( 4 * 1i ) * ...
        ( ones( numberOfPoints - 1, 1 ) ) / ( 2 * dx * dx );
    bC = 0.5 * ones( numberOfPoints, 1 ) - stepLength/( 4 * 1i ) *...
        ( potential( gridPointsC ) + ...
        ones( numberOfPoints, 1 ) / ( dx * dx ) );
else
% Crank-Nicholson solver based on built-in Octave sparse-matrix
tridiagonal routine
    onesVectorC = ones( numberOfPoints, 1 );
    leftMatrixRCs = speye( numberOfPoints ) + i * stepLength / 2 *...
        ( spdiags( potentialC, 0, numberOfPoints, numberOfPoints ) -...
        spdiags( [ onesVectorC, -2 * onesVectorC, onesVectorC ] /...
        ( 2 * dx * dx ), -1 : 1, numberOfPoints, numberOfPoints ) );
end
for loop = 1 : numberOfTimeSteps
    if enableMyTridiagonal
        wavefunctionC = MyTridiagonal( aC, bC, aC, wavefunctionC ) ...
```

```
                 - wavefunctionC;
     else
       wavefunctionC = 2 * ( leftMatrixRCs \ wavefunctionC ) - ...
           wavefunctionC;
     end
     if ( rem(loop,50) == 0 )
       plot( gridPointsC, abs( wavefunctionC ), 'g' );
     end
     drawnow
   end
```

In the above program, either the built-in Octave sparse-matrix equation solution routines or the tridiagonal matrix solver below can be employed by changing
enableMyTridiagonal from 1 to 0 (if column vectors are passed to **myTridiag-
onal( )** the return value is a column vector, whereas a row vector is returned for
row-vector parameters):

```
function outputVector = myTridiagonal( aLowerCodiagonal, ...
     aDiagonal, aUpperCodiagonal, aInputVector )
numberOfEquations = length( aDiagonal );
rowSize = size( aLowerCodiagonal );
if rowSize == 1
     outputVector = zeros( 1, numberOfEquations )
else
     outputVector = zeros( numberOfEquations, 1 );
end
for loop = 2 : numberOfEquations
     temporary = aLowerCodiagonal( loop - 1 ) / ...
         aDiagonal( loop - 1 );
     aDiagonal( loop ) = aDiagonal( loop ) - temporary * ...
         aUpperCodiagonal( loop - 1 );
     aInputVector( loop ) = aInputVector( loop ) - temporary * ...
         aInputVector( loop - 1 );
end
outputVector( numberOfEquations ) = aInputVector( ...
     numberOfEquations ) / aDiagonal( numberOfEquations );
for loop = numberOfEquations - 1 : -1 : 1
     outputVector( loop ) = ( aInputVector( loop ) - ...
         aUpperCodiagonal( loop ) * outputVector( loop + 1 ) ) /...
         aDiagonal( loop );
end
```

24.9   Finite-difference and finite-element procedures

The finite-difference method also yields the complete set of eigenvectors and
eigenfunctions of differential equations such as the Schrödinger or diffusion
equation. For example, the ground-state eigenvalue and eigenfunction are
obtained, for zero boundary conditions at the points x_0 and x_{N+1} each located

one point beyond the computational window boundary, by solving $\mathbf{H}\varphi_m = E_m\varphi_m$ according to

```
numberOfPoints = 100;
computationalWindowWidth = 20;
deltaX = computationalWindowWidth / ( numberOfPoints - 1);
leftEndPoint = -computationalWindowWidth / 2;
MRC = zeros( numberOfPoints, numberOfPoints );
xR = linspace( leftEndPoint, -leftEndPoint, numberOfPoints );
for ( loop = 1 : numberOfPoints )
    MRC(loop, loop) = 1 / deltaX^2 + potential( xR(loop) );
    if ( loop ~= numberOfPoints )
        MRC(loop, loop + 1) = -1 / ( 2 * deltaX^2 );
    end;
    if ( loop ~= 1 )
        MRC(loop, loop - 1) = -1 / ( 2 * deltaX^2 );
    end;
end
[ eigenVectorsC, eigenValues ] = eigs( MRC, 1, 'sa' );
plot( xR, eigenVectorsC * ...
    sign( eigenVectorsC(numberOfPoints / 2) ) );
```

Closely related to the finite-difference method is the *finite-element method*, in which the field is expressed as a linear superposition of localized basis functions with finite spatial extent. Since these basis functions overlap, the standard eigenvalue equation is instead replaced by a *generalized eigenvalue equation* of the form $\mathbf{H}|\varphi_k\rangle = \lambda_k\mathbf{M}|\varphi_k\rangle$, in which each eigenvalue multiplies a non-diagonal matrix.

The Galerkin derivation of the finite-element method, here specialized to a single dimension, follows from the observation that, if an eigensolution ψ of the Schrödinger equation is expanded as a superposition

$$\psi(x) = \sum_{j=1}^{N} a_j u_j(x) \tag{24.96}$$

of a complete set of localized but potentially overlapping functions $u_j(x)$, then for each value of i the identity

$$\int_{x_L}^{x_R} \left(-\frac{\hbar^2}{2m}u_i^*(x)\frac{d^2\psi(x)}{dx^2} + (V(x) - E)u_i^*(x)\psi(x) \right) dx = 0 \tag{24.97}$$

must be satisfied. Integrating this by parts,

$$\int_{x_L}^{x_R} u_i^*(x) \left(-\frac{\hbar^2}{2m}\frac{d^2\psi(x)}{dx^2} + (V(x) - E)\psi(x) \right) dx$$
$$= \int_{x_L}^{x_R} \left(\frac{\hbar^2}{2m}\frac{du_i^*(x)}{dx}\frac{d\psi(x)}{dx} + (V(x) - E)u_i^*(x)\psi(x) \right) dx - \frac{\hbar^2}{2m}u_i^*(x)\frac{d\psi(x)}{dx}\bigg|_{x_L}^{x_R} \tag{24.98}$$

For e.g. $u_i^*(x_R) = u_i^*(x_i) = 0$, the second, surface, term vanishes on the right-hand side of the above equation. Substituting Eq. (24.96) into Eq. (24.98) yields the matrix eigenvalue equation

$$S \begin{pmatrix} a_1 \\ a_2 \\ \vdots \end{pmatrix} = E \tilde{S} \begin{pmatrix} a_1 \\ a_2 \\ \vdots \end{pmatrix} \tag{24.99}$$

where the *structure matrix* S is given by

$$S_{ij} = \int_{x_L}^{x_R} \left(\frac{\hbar^2}{2m} \frac{du_i^*(x)}{dx} \frac{du_j(x)}{dx} + V(x) u_i^*(x) u_j(x) \right) dx \tag{24.100}$$

and

$$\tilde{S}_{ij} = \int_{x_L}^{x_R} u_i^*(x) u_j(x) dx \tag{24.101}$$

Consider now the *triangular basis functions*,

$$u_i(x) = \begin{cases} (x_{i+1} - x)/(x_{i+1} - x_i) & x_i < x < x_{i+1} \\ (x - x_{i-1})/(x_i - x_{i-1}) & x_{i-1} < x < x_i \\ 0 & \text{else} \end{cases} \tag{24.102}$$

where for the first function, $u_1(x)$, $x_0 = x_L$ coincides with the left endpoint of the computation interval and for the last function, $u_N(x)$, $x_{N+1} = x_R$ coincides with the right endpoint of the interval. Since the derivative of each basis function is $\pm 1/(\Delta x)^2$, adding the contributions, termed *element matrices*, of each interval to the structure matrices yields (again noting that $u_0 = u_{N+1} = 0$ from the zero boundary conditions), for four points located at $x_L + \Delta x, x_L + 2\Delta x, \ldots, x_L + 4\Delta x$ with $x_L + 5\Delta x = x_R$,

$$S = \sum_{i=1}^{5} S^{(i)} = \begin{pmatrix} \frac{\hbar^2}{2m\,\Delta x} + \int_{x_L}^{x_L+\Delta x} V(x) u_1 u_1\, dx & 0 & 0 & 0 \\ 0 & & 0 & 0 & 0 \\ 0 & & 0 & 0 & 0 \\ 0 & & 0 & 0 & 0 \end{pmatrix}$$

$$+ \begin{pmatrix} \frac{\hbar^2}{2m\,\Delta x} + \int_{x_L+\Delta x}^{x_L+2\Delta x} V(x) u_1 u_1\, dx & -\frac{\hbar^2}{2m\,\Delta x} + \int_{x_L+\Delta x}^{x_L+2\Delta x} V(x) u_1 u_2\, dx & 0 & 0 \\ -\frac{\hbar^2}{2m\,\Delta x} + \int_{x_L+\Delta x}^{x_L+2\Delta x} V(x) u_1 u_2\, dx & \frac{\hbar^2}{2m\,\Delta x} + \int_{x_L+\Delta x}^{x_L+2\Delta x} V(x) u_2 u_2\, dx & 0 & 0 \\ 0 & 0 & 0 & 0 \\ 0 & 0 & 0 & 0 \end{pmatrix} + \cdots$$

$$\tag{24.103}$$

together with

$$
\tilde{S} = \begin{pmatrix}
\dfrac{2\,\Delta x}{3} & \dfrac{\Delta x}{6} & 0 & 0 \\[2mm]
\dfrac{\Delta x}{6} & \dfrac{2\,\Delta x}{3} & \dfrac{\Delta x}{6} & 0 \\[2mm]
0 & \dfrac{\Delta x}{6} & \dfrac{2\,\Delta x}{3} & \dfrac{\Delta x}{6} \\[2mm]
0 & 0 & \dfrac{\Delta x}{6} & \dfrac{2\,\Delta x}{3}
\end{pmatrix} \tag{24.104}
$$

After the upper triangular parts of the matrices $\mathbf{S}$ and $\tilde{\mathbf{S}}$ have been evaluated the full matrices can be obtained through symmetrization (which doubles the diagonal elements). If the potential function remains constant, V_0, over a finite element, the integrals over the potential function in S equal V_0 times the corresponding elements of $\tilde{S}$. Approximating the potential function over each triangular element by its value at the center of the element and employing the generalized Octave eigenvalue solver for $\mathbf{A}\vec{v} = \lambda \mathbf{B}\vec{v}$, **eigs(A, B)**, yields, after summing the matrices in Eq. (24.103),

```
numberOfPoints = 100;
computationalWindowWidth = 20;
deltaX = computationalWindowWidth / ( numberOfPoints - 1 );
leftEndPoint = -computationalWindowWidth / 2;
SRC = zeros( numberOfPoints, numberOfPoints );
SBarRC = zeros( numberOfPoints, numberOfPoints );
% Coordinates of the center of each finite element
xR = linspace( leftEndPoint, -leftEndPoint, numberOfPoints );
% Boundary points are 0 (at x_L = -leftEndPoint - deltaX)
% and numberOfPoints + 1
for ( loop = 1 : numberOfPoints )
        SRC(loop, loop) = 1 / ( 2 * deltaX ) + ...
                potential( xR(loop) ) * deltaX / 3;
        SBarRC(loop, loop) = deltaX / 3;
        if ( loop ~= numberOfPoints )
                SRC(loop, loop + 1) = -1 / ( 2 * deltaX ) + ...
                        potential( xR(loop) ) * deltaX / 6;
                SBarRC(loop, loop + 1) = deltaX / 6;
        end
end
% Symmetrization
SRC = SRC + SRC';
SBarRC = SBarRC + SBarRC';
% Generalized eigenvalue routine
[ eigenVectorsC, eigenValues ] = eigs( SRC, SBarRC, 1, 'sa' );
plot( xR, eigenVectorsC * ...
        sign( eigenVectorsC(numberOfPoints / 2) ) );
```

Index

For EU product safety concerns, contact us at Calle de José Abascal, 56–1°,
28003 Madrid, Spain or eugpsr@cambridge.org.

www.ingramcontent.com/pod-product-compliance
Ingram Content Group UK Ltd.
Pitfield, Milton Keynes, MK11 3LW, UK
UKHW050216090126
466816UK00009B/85